An Introduction to Plasma Physics and its Space Applications, Volume 2

Basic equations and applications

An Introduction to Plasma Physics and its Space Applications, Volume 2

Basic equations and applications

Luis Conde
Universidad Politécnica de Madrid, Madrid, Spain

IOP Publishing, Bristol, UK

ISBN 978-0-7503-3543-0 (ebook)
ISBN 978-0-7503-3541-6 (print)
ISBN 978-0-7503-3544-7 (myPrint)
ISBN 978-0-7503-3542-3 (mobi)

DOI 10.1088/978-0-7503-3543-0

Version: 20200801

IOP ebooks

British Library Cataloguing-in-Publication Data: A catalogue record for this book is available from the British Library.

Published by IOP Publishing, wholly owned by The Institute of Physics, London

IOP Publishing, Temple Circus, Temple Way, Bristol, BS1 6HG, UK

US Office: IOP Publishing, Inc., 190 North Independence Mall West, Suite 601, Philadelphia, PA 19106, USA

To my wife

Contents

Preface ix

Acknowledgements x

Author biography xi

Acronyms xii

Symbols xiii

1 Introduction to the physical models for plasmas 1-1

2 Elements of plasma kinetic theory 2-1

2.1 The Boltzmann equation 2-3
2.2 Relaxation model for molecular collisions 2-7
2.3 The Boltzmann collision integral 2-8
2.3.1 Qualitative derivation 2-8
2.3.2 Approximations 2-11
2.3.3 The Maxwell–Boltzmann distribution 2-11
2.4 Commentaries and further reading 2-12
References 2-12

3 Hydrodynamic description of plasmas 3-1

3.1 The moments of Boltzmann equation 3-4
3.1.1 The equation of continuity 3-9
3.1.2 The momentum transport equation 3-10
3.1.3 The friction force 3-13
3.1.4 The energy transport equation 3-15
3.2 The hydrodynamic plasma transport equations 3-18
3.2.1 The closure of the transport equations 3-19
3.2.2 The cold and warm plasma models 3-20
3.2.3 The diffusion approximation 3-22
3.3 Electron and ion waves 3-23
3.3.1 Neutral gas sound waves 3-24
3.3.2 Ion waves 3-25
3.3.3 Electron waves 3-27
3.4 The Langmuir sheath 3-29
3.4.1 Bohm criterion 3-31

3.4.2 Child–Langmuir law 3-33
3.4.3 Space-charge neutralization 3-34
3.5 Commentaries and further reading 3-35
References 3-36

4 Introduction to electric discharges 4-1
4.1 DC electric discharges 4-2
4.1.1 Vacuum breakdown in DC fields 4-4
4.2 Microwave discharges 4-8
4.3 Breakdown in high frequency fields 4-9
4.4 Commentaries and further reading 4-11
References 4-11

5 The multipactor discharge 5-1
5.1 The electron multipactor discharge 5-2
5.2 The multipactor discharge model 5-4
5.3 Commentaries and further reading 5-10
References 5-11

6 Introduction to plasma propulsion in space 6-1
6.1 Basic orbital maneuvers 6-1
6.2 The Tsiolkovsky equation 6-3
6.3 Thrust and specific impulse 6-5
6.4 Electric propulsion principles 6-6
6.4.1 Efficiencies 6-9
6.5 Modified Tsiolkovsky equation 6-10
6.6 Classification of electric thrusters 6-13
6.6.1 Electrothermal 6-13
6.6.2 Electrostatic 6-15
6.6.3 Electromagnetic 6-19
6.7 Commentaries and further reading 6-20
References 6-20

Appendices

Appendix A A-1

Appendix B B-1

Preface

The physics of ionized gases and plasmas plays a relevant role in aerospace technology and requires that graduate students and professionals in the field understand their principles. For them, microwave electric discharges in telecommunication equipment testing or plasma propulsion of satellites are today familiar terms.

This book is the expanded version of the lectures and class notes I taught on this subject to students of aerospace engineering and physics and it is intended as a starting point. For this reason I left aside the conventional organization of plasma physics textbooks to make the subject more accessible to non-specialists.

In the first volume the fundamentals of plasma physics, the properties of molecular and atomic collisions and elementary plasma processes were introduced. This second volume is an introduction to the physical description of interacting particle systems where I tried do not leave the algebraic steps to obscure the physics and/or to be an additional difficulty for the reader. It starts from basic kinetic theory and the hydrodynamic transport equations for a plasma are derived next. The starting point is the velocity distribution function to make the subject more accesible to the beginner. The formal derivation of the kinetic plasma equation is discussed in appendix B. Finally, the readers will be introduced to microwave multipactor electric breakdown and the basis of in-space plasma propulsion.

This introductory book is limited in scope and missing relevant issues in plasma physics that I hope readers can address later. Each chapter concludes with a selection of recommended materials for further reading that are basically dictated by my personal choice. I apologize for the many books and papers that undoubtedly deserve the attention of readers and are inevitably not mentioned with this limiting criterion.

Finally, I made all efforts to make the text free of errors but this is a hopeless task. I will be grateful to readers who kindly inform me of misprints and errors, preferably by electronic mail.

Luis Conde
Madrid, November 2019
luis.conde@upm.es

Acknowledgements

I am grateful to my colleagues J M Donoso and J L Domenech-Garret who read the two volumes of this book and have made valuable commentaries and suggestions. I benefited from the helpful feedback and of early versions by S Tierno and J González and stimulating discussions with I M Tkachenko. I also appreciate the support and interest of Robert Trevelyan and David McDade from IOP Publishing. Finally, and most of all to my beloved wife for her patience while writing this book.

Author biography

Luis Conde

Luis Conde is Professor of Physics and the Polytechnic University of Madrid (Universidad Politcnica de Madrid, UPM). He received his PhD in Physics from the Open University of Madrid on Macromolecular Physics and has been with the Spanish National Research Council, the Complutense University of Madrid and he joined the UPM Faculty of Aerospace Engineering as a member of the Department of Applied Physics on1989. He has given lecture courses on plasma physics to master and undergraduate students of aerospace engineering and physics. He is Fellow of the European Physical Society and has authored/co-authored over 50 journal articles and 49 conference papers. In 2019 he received two patents (US 10172227B2 and EP3369294B1) of an innovative low electric power plasma thruster. His current interests are basic plasma physics, plasma diagnostics and plasma thrusters for space propulsion.

Acronyms

AC	alternating current.
AF-MPD	applied-field magnetoplasmadynamic thruster.
AIP	American Institute of Physics.
APS	American Physical Society.
AZ	anode zone.
BGK	Bathnagar–Gross–Krook relaxation model.
CEX	charge exchange collision.
CM	center of mass.
DC	discrete current.
ECR	electron cyclotron resonance.
CDS	cathode dark space.
EP	electric propulsion.
ESA	European Space Agency.
EWSK	east–west station keeping.
FDS	Faraday dark space.
FEEP	field emission electric propulsor.
GEO	geostationary Earth orbit.
GIE	gridded ion engine.
GOCE	gravity field and steady state ocean circulator Earth explorer.
GTO	geosynchronous transfer orbit.
HEMPT	highly efficient multistage plasma thruster.
HET	Hall effect thruster.
IOP	Institute of Physics.
IV	current–voltage curve of a DC electric discharge.
LEO	low Earth orbit.
LTE	local thermal equilibrium.
MB	Maxwell–Boltzmann.
MEO	medium Earth orbit.
mfp	mean free path.
MHD	magnetohydrodynamic equations.
MPD	magnetoplasmadynamic thruster.
NASA	National Aeronautics and Space Administration.
NG	negative glow.
NSSK	north–south station keeping.
PC	positive column.
PLE	partial local equilibrium.
RF	radio frequency.
RST	radioisotope thermoelectric generator.
SF-MPD	self-field magnetoplasmadynamic thruster.
SEY	secondary emission yield.
STP	standard temperature and pressure.
UV	ultraviolet electromagnetic radiation.

Symbols

Å	Ångstrom (length) 1 Å = 10^{-8} cm.
amu	atomic mass unit 1.66×10^{-27} kg.
α_g	ionization degree of a gas.
α_w	*specific power*; electric power to weight of power equipment in W kg^{-1}.
c_{is}	ion sound speed.
e	electron charge $e = 1.6021 \times 10^{-19}$ C.
$e_{i\alpha}$	internal energy per particle $\alpha = i, e, a$.
E_I, E_{II}	ionization energy of a neutral atom (*I* first level, *II* second, etc).
ε_o	permittivity of a vacuum $\varepsilon_o = 8.8542 \times 10^{-12}$ F m^{-1}.
eV	electron volt 1 eV = 1.602×10^{-19} J.
η_e	electric efficiency of a thruster.
η_m	mass utilization efficiency of a thruster.
η_T	total efficiency $\eta_T = \eta_m \times \eta_e$
f_{pe}	electron plasma frequency $f_{pe} = \omega_{pe}/2\pi$ in Hz.
$f_\alpha(\boldsymbol{r}, \boldsymbol{v}, t)$	non-equilibrium velocity distribution function.
$f_{mb}(\boldsymbol{v})$	Maxwell–Boltzmann velocity distribution function.
f_{pi}	ion plasma frequency $f_{pe} = \omega_{pi}/2\pi$ in Hz.
F	thrust $F = \dot{m}_p v_{ex}$.
$\phi_c(r)$	Coulomb potential $\phi_c(r) = q/(4\pi\varepsilon_o r)$.
$g_{mb}(E)$	Maxwell–Boltzmann energy distribution function.
g_o	Standard Earth's gravity $g_o = 9.8$ m s^{-1}.
Hz	Hertz (1 cycle/second).
GHz	giga Hertz (10^9 Hz).
I_{sp}	specific impulse in seconds.
I_t	total impulse in seconds.
$\boldsymbol{k}$	wave number $k = 2\pi/\lambda$.
k_B	Boltzmann's constant $k_B = 1.3805 \times 10^{-23}$ J K^{-1}.
K	Kelvin degree (temperature).
kHz	kHz (10^3 Hz).
keV	kilo electron-volts, 10^3 eV.
L_s	characteristic length scale (its meaning depends on the context).
λ	wavelength.
$\lambda_{D\alpha}$	Debye length for $\alpha = e, i$ charged particles.
λ_c	mean free path for collisions with neutral atoms.
mbar	millibar, gas pressure unit 1 atmosphere = 1013.25 mbar.
MHz	mega Hertz (10^6 Hz).
m_e	electron mass $m_e = 9.1091 \times 10^{-23}$ J K^{-1}.
MeV	mega electron-volt, 10^6 eV.
mN	milliNewton (10^{-3} N), unit of force.
N	Newton (force unit).
n_a	number density of neutrals. Number of neutral atoms or molecules by unit volume.
$N_{D\alpha}$	plasma parameter for $\alpha = e, i$ charged particles.
n_e	number density of electrons. Number of electrons by unit volume.
n_i	number density of ions. Number of ions by unit volume.
ν_c, ν_m	collision frequencies with neutral atoms.

ν_I	electron impact ionization frequency.
ω	wave angular frequency.
ω_{pe}	electron plasma frequency $\omega_{pe} = 2\pi\, f_{pe}$ in radians per second.
ω_{pi}	electron plasma frequency $\omega_{pi} = 2\pi\, f_{pi}$ in radians per second.
p_a	pressure of the neutral gas.
P_b	kinetic energy power (in watts) $P_b = \dot{m}_p\, v_{ex}^2/2$.
τ_b	burning time of a thruster.
τ_c	time elapsed between two successive atomic or molecular collisions.
τ_r	characteristic relaxation time.
τ_s	characteristic time scale (its meaning depends on the context).
τ_{ia}	time elapsed between two successive ion/atom collisions.
τ_{ea}	time elapsed between two successive ion/electron collisions.
$\bar{v}$	mean thermal speed $(8\, k_B\, T/\pi\, m)^{1/2}$.
v_{ex}	propellant exhaust velocity.
$v_f = \omega/k$	wave phase velocity.
$v_g = \partial\omega/\partial\boldsymbol{k}$	wave group velocity.
v_{th}	thermal speed $\sqrt{2\, k_B T_\alpha/m}$.

IOP Publishing

An Introduction to Plasma Physics and its Space Applications,
Volume 2
Basic equations and applications
Luis Conde

Chapter 1

Introduction to the physical models for plasmas

The fundamental scheme of theories currently employed for the physical description of large number of interacting particles is discussed in first place. The plasma kinetic theory and statistical mechanics consider a microscopic approach where electrons, ions, neutral atoms and their elementary processes at atomic and molecular level are considered. In the opposite limit are macroscopic fluid models where plasma is described as homogeneous continuum media and transport coefficients such as viscosity or thermal conductivity account for its specific properties.

The plasma state of matter was introduced in section 2.3 of volume 1 as a system composed of a large number of ions, electrons and neutral atoms. The basic physical parameters of *ideal plasmas* were introduced in chapter 4 of volume 1 on the basis of the statistical description of electrically neutral gases in thermal equilibrium. Specifically, the Maxwell–Boltzmann distribution function, equations (3.5) or (3.7) in volume 1, was used to calculate the Debye length that governs the spatial fluctuations of the plasma electric field. This statistical description of gases and plasmas will be generalized in the following to account for time-dependent physical processes and/or transport phenomena where the equilibrium Maxwell–Boltzmann distribution function is no longer valid.

The basic structure of physical theories currently employed to describe a statistical ensemble of interacting particles are outlined in the diagram of figure 1.1. The *microscopic* approaches (statistical mechanics and kinetic theory) make use of statistical averages that take into account specific properties of atoms or molecules. The macroscopic properties that can be measured in the laboratory are calculated from such averages. Alternatively, thermodynamics or fluid dynamics are

doi:10.1088/978-0-7503-3543-0ch1

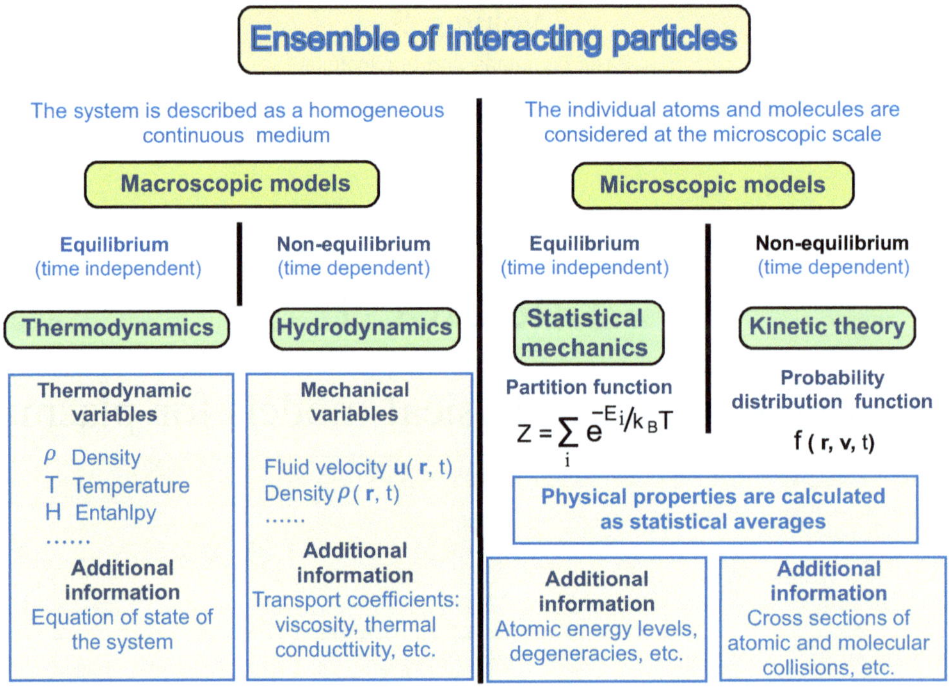

Figure 1.1. Scheme with the more relevant physical theories describing an statistical ensemble of interacting particles. The *additional information* is coefficients and/or equations that cannot be directly derived and need to be provided for the mathematical closure of the physical model.

macroscopic approaches which consider the physical system as a continuum medium, disregarding specific details of its constituent particles at the microscopic level.

Thermodynamic and statistical physics essentially apply to *equilibrium states* of condensed matter, as indicated in figure 1.1. Both formulations are valid when the temporal evolution of the system is much slower than the time scales involved in the physical processes at atomic and molecular levels that relax fluctuations to the equilibrium state. Thus, *time* does not explicitly appear in the equations of thermodynamic and statistical physics because the system is considered always in an equilibrium state and/or very close to it slowly evolving in time.

In contrast, the physical description of *non-equilibrium* macroscopic properties explicitly considers the time evolution of the system. The hydrodynamics and kinetic theory are more involved and concern *non-equilibrium* states where the physical system explicitly changes in time, as shown in the scheme of figure 1.1. The fluid transport equations for a plasma can be derived by considering the medium as a continuum using the mass, momentum and energy conservation principles, together with the Maxwell equations (2.6) of volume 1 for the macroscopic electromagnetic fields. However, the mathematical closure of the problem needs additional physical magnitudes, such as the thermal conductivity or viscosity (called transport coefficients) that are introduced as additional phenomenological expressions.

The starting point of kinetic theory is the *velocity distribution function* $f_\alpha(\boldsymbol{r}, \boldsymbol{v}, t)$ for each plasma species (α = e, i, a, electrons, ions and neutral atoms) that will be introduced in chapter 2. In the general case, it depends on the velocity $\boldsymbol{v}$, position $\boldsymbol{r}$ and also evolves in time. It is proportional to the probability of finding at instant t one particle at point $\boldsymbol{r}$ moving with $\boldsymbol{v}$ velocity. As we shall see, this distribution can be calculated as the solution of the integro-differential *kinetic* or *Boltzmann equation* which accounts for the average forces acting on plasma particles at the microscopic level.

The time-dependent macroscopic properties, such as energy, pressure or average speed are calculated as statistical averages using $f_\alpha(\boldsymbol{r}, \boldsymbol{v}, t)$ and appropriate operators for these physical magnitudes. This procedure was previously employed in section 3.3 for neutral gases using the equilibrium distributions (3.5) and (3.7). This Maxwell–Boltzmann distribution for neutral atoms is recovered from the Boltzmann integral in the limit corresponding to the thermodynamic equilibrium state.

The complex statistical approach of kinetic theory is appealing when the relevant collisional mean free path λ_c and/or collision time $\tau_c = 1/\nu_c$ can be compared to the characteristic length L_s or the time scale τ_s of the problem under consideration. Equivalently, this can also be when $f_\alpha(\boldsymbol{r}, \boldsymbol{v}, t)$ appreciably changes along the typical length L_s and/or on plasma time scale τ_s of the system. However, the diagram of figure 1.1 shows the mathematical closure of kinetic formulation also requires *supplementary information*, such as the cross sections of particle collisions introduced in chapters 5 and 6 of volume 1.

In the *hydrodynamic* approach plasmas are regarded as *macroscopic continuum media* where a huge number of particles are contained within small volumes with typical size l much smaller $L_s \gg l \gtrsim \lambda_D$ than the length scale L_s of the system, but larger than the Debye length. The plasma hydrodynamic equations consider three mutually inter-penetrating fluids of neutral atoms, electrons and ions which motions are coupled by electromagnetic interaction and particle collisions and are more involved than those of ordinary fluids. In essence, the more tractable fluid transport equations apply when the relevant mean free path λ_c for collisions is $L_s \gg l \gg \lambda_c$ much smaller than characteristic length L_s of the problem and then, the plasma is dominated by collisions.

As discussed in section 4.2 of volume 1, this hydrodynamic approach is valid when collisions between particles relax the random fluctuations on a fast time scale to a *local* probability distribution function, as was the case of LTE plasmas discussed in section 4.3 of volume 1. Equivalently, when the distribution function $f_\alpha(\boldsymbol{r}, \boldsymbol{v}, t)$ changes smoothly along typical distances $l \gg \lambda_c$ much longer than the collisional mean free paths λ_c. As we shall see in chapter 3, in this case, the hydrodynamic equations can be alternatively derived from kinetic theory assuming an unspecified velocity distribution function.

Finally, the collisions between plasma species connect both hydrodynamic and kinetic formulations of plasma state. The energy transfer between particle species by elastic collisions, the ionization and recombination rates, etc, are incorporated in the kinetic equations by means of *collision operators* that account for the changes introduced in $f_\alpha(\boldsymbol{r}, \boldsymbol{v}, t)$ by elementary processes. We will see in chapter 3 that their macroscopic average will provide us with expressions for the energy and momentum exchange that are needed in the three-fluids hydrodynamic description to account for the mechanical coupling between ion and electrons fluids, etc.

The basis of kinetic description of plasmas is introduced in chapter 2 where this *microscopic* approach will show the connection between the elementary processes and the non-equilibrium probability distribution functions discussed in chapters 5 and 6 of volume 1. The connection between the statistical averages and the individual particle properties is a complex theory and in appendix B is outlined a more rigorous derivation of the Boltzmann or kinetic equation, which is the basis of the kinetic description of chapter 2 where we will follow a more intuitive approach.

The hydrodynamic approach will be introduced in chapter 3 where the fluid equations are derived from the kinetic description. As we shall see, macroscopic transport coefficient such as plasma thermal conductivity or viscosity is also connected with collisional processes at the microscopic level. The fluid equations are derived from statistical averages of the mechanical magnitudes of interest such as energy and momentum. However, we shall see some physical assumptions are required for the closure of macroscopic transport equations.

IOP Publishing

Chapter 2

Elements of plasma kinetic theory

In this chapter we introduce the advanced *kinetic* description of non-equilibrium plasmas that generalizes the statistical concepts for neutral gases in equilibrium introduced in chapter 3 of volume 1. The Boltzmann equation governs the time evolution of the probability distribution function of electrons, ions and neutral atoms. The evolution introduced by collisions at the microscopic level is discussed on the basis of the simple BGK relaxation model and the Boltzmann collision integral.

The plasma kinetic theory considers the plasma as an ensemble of interacting electrically charged particles and neutral atoms or molecules. However, we will consider in the following only one kind of particle to focus on basic concepts that will be later extended to all plasma species.

The dynamical state of a particle is determined in classical mechanics by its position $\boldsymbol{r}$ and velocity $\boldsymbol{v}$ (or momentum $\boldsymbol{p}$) at any given time. Then, we can represent a system of N particles by a collection of pairs $(\boldsymbol{r}_i, \boldsymbol{v}_i)$ corresponding to the $i = 1, 2, \ldots N$ particles, equivalent to N points distributed into a six-dimensional *phase space* $(\boldsymbol{r}, \boldsymbol{v}) = (r_x, r_y, r_z, v_x, v_y, v_z)$. The velocity $\boldsymbol{v}$ and position $\boldsymbol{r}$ are independent variables that are respectively called *velocity space* and *geometry space*.

For a large number of particles it is statistically meaningful to consider the number of particles (points) dN contained within a small volume element $d^3r\, d^3v = dx\, dy\, dz\, dv_x\, dv_y\, dv_z$ in six dimensions of the phase space located at point $(\boldsymbol{r}, \boldsymbol{v})$. Equivalently, dN is also the number of particles within the volume located between $\boldsymbol{r}$ and $\boldsymbol{r} + d\boldsymbol{r}$ in the geometry space with velocities between $\boldsymbol{v}$ and $\boldsymbol{v} + d\boldsymbol{v}$. We can also write $d^3v\, d^3r = d\boldsymbol{v}\, d\boldsymbol{r}$ to simplify the mathematical notation and,

doi:10.1088/978-0-7503-3543-0ch2

$$dN = f(\boldsymbol{v}, \boldsymbol{r}, t)\, d^3v\, d^3r = f(\boldsymbol{v}, \boldsymbol{r}, t)\, d\boldsymbol{v}\, d\boldsymbol{r} \tag{2.1}$$

where $f(\boldsymbol{v}, \boldsymbol{r}, t)$ is the *velocity distribution function*[1]. Within an elementary volume (d^3r) of the geometrical space, the number density of particles dn is,

$$dn = \frac{dN}{dr^3} = f(\boldsymbol{v}, \boldsymbol{r}, t)\, d^3v \quad \text{also,} \quad \delta N = \frac{dN}{dr^3\, dv^3} = f(\boldsymbol{v}, \boldsymbol{r}, t)$$

Then $n(\boldsymbol{r}, t)$ is the particle density function, whereas δN represents the number of points contained within the volume $d^3r\, d^3v$ of the phase space. Therefore the function,

$$P(\boldsymbol{v}, \boldsymbol{r}, t) = \frac{\delta N}{N} = \frac{1}{N} \times \frac{dN}{dr^3\, dv^3} = \frac{1}{N} f(\boldsymbol{v}, \boldsymbol{r}, t) \tag{2.2}$$

gives the probability $P(\boldsymbol{v}, \boldsymbol{r}, t)$ of finding one particle with velocity in the range $(\boldsymbol{v}, \boldsymbol{v} + d\boldsymbol{v})$ and position between $(\boldsymbol{r}, \boldsymbol{r} + d\boldsymbol{r})$ in the geometry space. This *probability distribution function* evolves in time, and is normalized to the unity,

$$\int_{-\infty}^{+\infty} \int_{-\infty}^{+\infty} P(\boldsymbol{v}, \boldsymbol{r}, t)\, d\boldsymbol{v}\, d\boldsymbol{r} = 1 \quad \text{and equivalently,}$$
$$\int_{-\infty}^{+\infty} \int_{-\infty}^{+\infty} f(\boldsymbol{v}, \boldsymbol{r}, t)\, d\boldsymbol{v}\, d\boldsymbol{r} = N \tag{2.3}$$

where N is the total number of particles in the system. The integrals are extended to all velocities and positions possible to cover the entire $(\boldsymbol{r}, \boldsymbol{v})$ phase space.

The macroscopic energy of the system is finite, so the sum of the kinetic energies of all particles must be bounded. To achieve this, the probability of finding a rapidly moving particle must decrease as its speed $v = |\boldsymbol{v}|$ increases at all points $\boldsymbol{r}$ of the geometry space. Otherwise, the macroscopic energy of the system would grow without limit. Mathematically, this means that the function $f(\boldsymbol{v}, \boldsymbol{r}, t)$ tends to zero when the speed $|\boldsymbol{v}|$ becomes infinitely large at all positions of $\boldsymbol{r}$ of the geometry space.

The *stationary* $f(\boldsymbol{v}, \boldsymbol{r})$ velocity distribution function has $\partial f / \partial t = 0$. When it is independent of $\boldsymbol{r}$ as $f(\boldsymbol{v}, t)$ it is said to be *homogeneous*, and *inhomogeneous* if it is not uniform in the geometry space. The distribution is *isotropic* when $f(|\boldsymbol{v}|, \boldsymbol{r}, t)$ is independent of the direction of the velocity $\boldsymbol{v}$ and it is called *anisotropic* otherwise.

The equilibrium Maxwell–Boltzmann distribution, equation (3.5) of volume 1, is a *stationary*, *homogeneous* and *isotropic* probability distribution function. Its properties reflect the physical characteristics of the thermodynamic equilibrium state at microscopic level; the molecules are equally distributed in space and their motions have no preferred direction since all are equally likely.

As in section 3.3 in volume 1, we can introduce the macroscopic physical magnitudes as averages of the distribution function $f(\boldsymbol{v}, \boldsymbol{r}, t)$ over the phase space $(\boldsymbol{r}, \boldsymbol{v})$. Since $\delta N = f(\boldsymbol{v}, \boldsymbol{r}, t)\, d\boldsymbol{v}\, d\boldsymbol{r}$ gives the number of points inside the elementary

[1] It will be called also *distribution function* in the following for short.

volume ($d^3r\ d^3v$) and $n(\boldsymbol{r}, t)$ is the number density of particles within d^3r, integration gives

$$n(\boldsymbol{r}, t) = \int_{-\infty}^{+\infty} f(\boldsymbol{v}, \boldsymbol{r}, t)\, d\boldsymbol{v} \text{ and also, } N = \int_{-\infty}^{+\infty} n(\boldsymbol{r}, t)\, d\boldsymbol{r} \tag{2.4}$$

We can also introduce the average or *macroscopic flow velocity*,

$$\boldsymbol{u}(\boldsymbol{r}, t) = \frac{\int_{-\infty}^{+\infty} \boldsymbol{v}\, f(\boldsymbol{v}, \boldsymbol{r}, t)\, d\boldsymbol{v}}{\int_{-\infty}^{+\infty} f(\boldsymbol{v}, \boldsymbol{r}, t)\, d\boldsymbol{v}} = \frac{1}{n_\alpha(\boldsymbol{r}, t)} \int_{-\infty}^{+\infty} \boldsymbol{v}\, f(\boldsymbol{v}, \boldsymbol{r}, t)\, d\boldsymbol{v} \tag{2.5}$$

which gives the *velocity field* $\boldsymbol{u}(\boldsymbol{r}, t)$ and the *flux of particles* as,

$$\boldsymbol{\Gamma}(\boldsymbol{r}, t) = \int_{-\infty}^{+\infty} \boldsymbol{v} f(\boldsymbol{v}, \boldsymbol{r}, t) d\boldsymbol{v} \tag{2.6}$$

These averages generalize concepts previously introduced in volume 1 section 3.2 for the Maxwellian distribution equations (3.5) or (3.7) of a neutral gas.

The determination of the distribution function is a central problem of kinetic theory since macroscopic physical properties are calculated as statistical averages. The evolution of $f(\boldsymbol{v}, \boldsymbol{r}, t)$ is governed by the *Boltzmann* or *kinetic equation* which gives the temporal and spatial evolution of the distribution function. However, the *probability distribution function* can be used instead and $dP = P(\boldsymbol{v}, \boldsymbol{r}, t)\, d\boldsymbol{v}\, d\boldsymbol{r}$ gives the probability of finding one particle within the elementary volume ($d^3r\ d^3v$) of the phase space.

2.1 The Boltzmann equation

We are now in a position to generalize the concepts introduced in the previous section for a system composed of several different kinds of particles. In the kinetic description the N_α plasma particles, where $\alpha = e, i, a$ labels electrons, ions and neutral atoms, these are considered as points with mass m_α and electric charge q_α (with $q_a = 0$) distributed into the *phase space* ($\boldsymbol{r}, \boldsymbol{v}$). Each plasma species has its own velocity distribution function $f_\alpha(\boldsymbol{v}, \boldsymbol{r}, t)$ that—as we shall see—is coupled by the elementary processes (collisions) between them at the microscopic level.

We will introduce here the Boltzmann equation by means of a somewhat simple and intuitive approach, starting from the definition (2.2) of the probability distribution function. The rigorous formulation in connection to the positions and velocities of point-like charged plasma particles and the derivation of the Boltzmann (or kinetic) equation is outlined[2] in appendix B.

[2] A more rigorous derivation of plasma kinetic equations (B.15) or (B.16) is in appendix B connecting the point-like distribution of particles in phase space with the probability distribution (2.2). The function $f_\alpha(\boldsymbol{r}, v, t)$ results from the average over an intermediate length scale L_c much longer than the interparticle distance $d_o = n_o^{-1/3} \ll L_c \lesssim \lambda_D$ but below the Debye length and velocities $v_c = L_c/T_\alpha \lesssim \lambda_D \times f_{p,\alpha}$ where $f_{p,\alpha} = 1/T_\alpha$ is the plasma frequency which introduces the faster time of response of the plasma.

The distribution function $f_\alpha(v, r, t)$ depends on seven independent variables[3] and using equation (A.3) its total time derivative is,

$$\frac{Df_\alpha}{Dt} = \frac{\partial f_\alpha}{\partial t} + \sum_i \left(\frac{\partial f_\alpha}{\partial x_i} \frac{dx_i}{dt} \right) + \sum_i \left(\frac{\partial f_\alpha}{\partial v_i} \frac{dv_i}{dt} \right)$$

that may be cast (equation A.4) as,

$$\frac{Df_\alpha}{Dt} = \frac{\partial f_\alpha}{\partial t} + \boldsymbol{v} \cdot \nabla_\mathrm{r} f_\alpha + \boldsymbol{a} \cdot \nabla_\mathrm{v} f_\alpha$$

In this equation,

$$\nabla_\mathrm{r} \equiv \left(\frac{\partial}{\partial x} \boldsymbol{i} + \frac{\partial}{\partial y} \boldsymbol{j} + \frac{\partial}{\partial z} \boldsymbol{k} \right) \text{ and, } \nabla_\mathrm{v} \equiv \left(\frac{\partial}{\partial v_x} \boldsymbol{i} + \frac{\partial}{\partial v_y} \boldsymbol{j} + \frac{\partial}{\partial v_z} \boldsymbol{k} \right)$$

respectively, are the nabla operators, ∇_r in the geometry space and ∇_v in the velocity space where $(\boldsymbol{i}, \boldsymbol{j}, \boldsymbol{k})$ are constant unit vectors. In order to derive the Boltzmann equation for the distribution $f_\alpha(\boldsymbol{r}, v, t)$ we recast the previous expression introducing the acceleration $\boldsymbol{a}$ or force by unit of mass as,

$$\boldsymbol{a} = \frac{\boldsymbol{F}}{m_\alpha} = \boldsymbol{f}_{g\alpha} + \frac{q_\alpha}{m_\alpha}(\boldsymbol{E} + v_\alpha \times \boldsymbol{B})$$

Here $\boldsymbol{f}_{g\alpha}$ represents other non-electromagnetic forces acting on the α particles[4] with electric charge q_α. The local electromagnetic fields $\boldsymbol{E}(\boldsymbol{r}, t)$ and $\boldsymbol{B}(\boldsymbol{r}, t)$ are spatially averaged[5] over distances $L \lesssim \lambda_D$ to preserve the plasma collective effects. We can write the time evolution equation,

$$\frac{Df_\alpha}{Dt} = \frac{\partial f_\alpha}{\partial t} + \boldsymbol{v} \cdot \nabla_\mathrm{r} f_\alpha + \left[\boldsymbol{f}_{g\alpha} + \frac{q_\alpha}{m_\alpha}(\boldsymbol{E} + \boldsymbol{v} \times \boldsymbol{B}) \right] \cdot \nabla_\mathrm{v} f_\alpha \tag{2.7}$$

When the non-equilibrium velocity distribution function $f_\alpha(\boldsymbol{r}, v, t)$ eventually remains unaltered in time we have $Df_\alpha/Dt = 0$. In this case equation (2.7) is denominated the *Vlasov equation* or *collisionless Boltzmann equation* and the velocity distribution function is also a constant of motion in the dynamical evolution of the system.

However, the plasma particles exchange energy and momentum in collisional processes which produce changes in the velocity distribution function of ions, electrons and neutral atoms. Collisions alter number of particles (points) initially located in the phase space volume $(d^3r\ d^3v)$ centered at $(\boldsymbol{r}, v)$ because their velocities and/or positions can change. Then, particles move to other points of the phase space

[3] The derivation of this scalar function DP_α/Dt and the connection between Lagrangian and Eulerian velocities are discussed in appendix A.

[4] The Boltzmann equation also applies to neutral gas particles setting the electric charge $q_\alpha = 0$, as they are not affected for the electromagnetic fields but are coupled by collisions with the electron and ion distribution functions.

[5] In appendix B it is discussed that these electromagnetic fields acting over the individual charge q_α result from a *spatial average* over typical lengths $L \lesssim \lambda_D$ of effective electric charge shielding.

after colliding because their initial velocities and positions are altered on a time scale governed by collisions. In the general case, molecular encounters alter the velocity distribution function $Df_\alpha/Dt \neq 0$ and we can write,

$$\frac{Df_\alpha}{Dt} = C(f_\alpha) \text{ where, } C(f_\alpha) = \left(\frac{\delta f_\alpha}{\delta t}\right)_{\text{col}}$$

represents the change in the velocity distribution f_α originated by collisional processes with the other plasma species. The rate of change also involves their distributions and,

$$C(f_\alpha) = \sum_\beta C_{\alpha,\beta}(f_\alpha, f_\beta) \tag{2.8}$$

where $C_{\alpha,\beta}(f_\alpha, f_\beta)$ represents the change in f_α by collisions of α particles with all other β species. At this point we do not have an explicit expression available for $C(f_\alpha)$ coupling the probability distribution of two plasma species, that will depend on the collisional cross sections, particle densities, etc, so that we will leave this point for later.

Finally, the *Boltzmann equation* for the α-particle distribution function is,

$$\frac{\partial f_\alpha}{\partial t} + \boldsymbol{v} \cdot \nabla_{\text{r}} f_\alpha + \left[\boldsymbol{f}_{g\alpha} + \frac{q_\alpha}{m_\alpha}(\boldsymbol{E} + \boldsymbol{v} \times \boldsymbol{B})\right] \cdot \nabla_{\text{v}} f_\alpha = C(f_\alpha) \tag{2.9}$$

where the *collision term* $C(f_\alpha)$ is still undefined and accounts for the dynamical evolution introduced by the encounters at the atomic and molecular level, averaged over an effective electric field shielding length[6] λ_D.

Equation (2.9) can be seen as a continuity equation for the velocity distribution function in the phase space $(\boldsymbol{r}, \boldsymbol{v})$. First, we use $\boldsymbol{v} \cdot \nabla_{\text{r}} f_\alpha = \nabla_{\text{r}} \cdot (\boldsymbol{v} f_\alpha)$ and also

$$\boldsymbol{a} \cdot \nabla_{\text{v}} f_\alpha = \left[\boldsymbol{f}_{g\alpha} + \frac{q_\alpha}{m_\alpha}(\boldsymbol{E} + \boldsymbol{v} \times \boldsymbol{B})\right] \cdot \nabla_{\text{v}} f_\alpha = \nabla_{\text{v}} \cdot (\boldsymbol{a} f_\alpha)$$

The electromagnetic fields $\boldsymbol{E}(\boldsymbol{r}, t)$ and $\boldsymbol{B}(\boldsymbol{r}, t)$ as well as $\nabla_{\text{v}} \cdot \boldsymbol{f}_{g\alpha} = 0$ are independent of $\boldsymbol{v}$, so we can also make use of,

$$\nabla_{\mathbf{v}} \cdot (\boldsymbol{v} \wedge \boldsymbol{B}) = \boldsymbol{B} \cdot \cancelto{0}{(\nabla_{\mathbf{v}} \wedge \boldsymbol{v})} + \boldsymbol{v} \cdot (\nabla_{\mathbf{v}} \wedge \boldsymbol{B})$$

Therefore, the *Boltzmann equation* (2.9) is equivalent to,

$$\frac{\partial f_\alpha}{\partial t} + \nabla_{\text{r}} \cdot (\boldsymbol{v} f_\alpha) + \nabla_{\text{v}} \cdot (\boldsymbol{a} f_\alpha) = C(f_\alpha) \tag{2.10}$$

where $C(f_\alpha)$ is the sum (2.8) of all changes experienced by the distribution function for the α species by collisions. Equation (2.10) has the implicit assumption that the

[6] Strictly speaking, the Debye length introduced in section 4.5 of volume 1 is only valid for Maxwellian electrons and ions. This is not the general case, so λ_D can be considered here as an *effective electric shielding* scale length.

temporal scales of all collisional processes considered in $C(f_\alpha)$ are much faster than the time evolution of the distribution function.

Finally, introducing a generalized coordinate $\boldsymbol{R} = (\boldsymbol{r}, \boldsymbol{v})$ in the phase space and its time derivative,

$$\boldsymbol{V} = (\boldsymbol{v}, \boldsymbol{a}) = \left(\boldsymbol{v} = \frac{d\boldsymbol{r}}{dt}, \boldsymbol{a} = \frac{d\boldsymbol{v}}{dt}\right)$$

Expression (2.10) can be cast into a *generalized continuity equation* for the distribution f_α in the phase space as,

$$\frac{\partial f_\alpha}{\partial t} + \nabla_{\boldsymbol{R}} \cdot (\boldsymbol{V} f_\alpha) = C(f_\alpha) \qquad (2.11)$$

Then, $C(f_\alpha)$ can be considered as a the source/sink term and the vector $\boldsymbol{q} = f_\alpha \boldsymbol{V}$ is the flux of $f_\alpha(\boldsymbol{R}, t) = f_\alpha(\boldsymbol{r}, \boldsymbol{v}, t)$. The collision operator $C(f_\alpha) = (\delta f_\alpha / \delta t)_{\text{col}}$ accounts for points (particles) added and/or removed from the volume $(d^3r\, d^3v)$ of the phase space.

The plasma elementary processes introduced in chapters 5 and 6 of volume 1 are responsible for the term $C_s(f_\alpha)$ in the Boltzmann equation (2.9) or (2.9) which introduces the collisional changes in the probability density during the evolution of the system. In *collisionless plasmas* $C(f_\alpha) = 0$ and the probability density $f_\alpha(\boldsymbol{v}, \boldsymbol{r}, t)$ is a conserved quantity in the dynamic evolution of the system.

The Boltzmann equations (2.9) or (2.9) needs to be complemented with the Maxwell equations for the electromagnetic fields $\boldsymbol{E}(\boldsymbol{r}, t)$ and $\boldsymbol{B}(\boldsymbol{r}, t)$ in the plasma. These fields result from the coupling of plasma charged particles and the externally applied or self-generated fields. The velocity distribution function $f_\alpha(\boldsymbol{v}, \boldsymbol{r}, t)$ can be used to calculate the average plasma density of charged particles ρ_c and the transported current density $\boldsymbol{J}_c$ as,

$$\rho_c = \sum_\alpha \rho_{c\,\alpha} = \sum_\alpha q_\alpha\, n_\alpha(\boldsymbol{r}, t) = \sum_\alpha q_\alpha \int_{-\infty}^{+\infty} f_\alpha(\boldsymbol{v}, \boldsymbol{r}, t)\, d\boldsymbol{v} \qquad (2.12)$$

and,

$$\boldsymbol{J}_c = \sum_\alpha \boldsymbol{J}_{e\alpha} = \sum_\alpha q_\alpha \int_{-\infty}^{+\infty} \boldsymbol{v}\, f_\alpha(\boldsymbol{v}, \boldsymbol{r}, t)\, d\boldsymbol{v} \qquad (2.13)$$

that should be introduced in the Maxwell equations to self-consistently calculate the electromagnetic fields $\boldsymbol{E}(\boldsymbol{r}, t)$ and $\boldsymbol{B}(\boldsymbol{r}, t)$.

From a formal point of view, and lacking an explicit formulation for the collisional term $C_s(f_\alpha)$, the kinetic approach to plasma physics may be understood as the closure of the Maxwell equations for $\boldsymbol{E}(\boldsymbol{r}, t)$ and $\boldsymbol{B}(\boldsymbol{r}, t)$ using Boltzmann equation (2.10).

When the distribution functions $f_\alpha(\boldsymbol{v}, \boldsymbol{r}, t)$ can be obtained, the physical macroscopic magnitudes, such as the local particle density $n_\alpha(\boldsymbol{r}, t)$ in equation (2.4) or the velocity field $\boldsymbol{u}_\alpha(\boldsymbol{r}, t)$ in equation (2.5) for ions, electrons and neutral atoms can be calculated. This approach was previously used in volume 1 sections 3.3 and 3.4 using the equilibrium Maxwell–Boltzmann distributions of equations (3.5) or (3.7).

However, to obtain explicit solutions of the Boltzmann equation is difficult. The generalization of this kinetic scheme to non-equilibrium plasmas is by far much more complex than for a neutral gas in equilibrium. Strictly speaking, we need to model the collision term $C_s(f_\alpha)$ accounting for the specific characteristics of collisional processes discussed in chapter 6 of volume 1.

The limited *relaxation model* introduced in the next section is a simple approach to introduce the effect of short-range collisions and next we will discuss the more involved *Boltzmann collision integral* valid for short-range binary collisions in low pressure gases and weakly ionized plasmas. The long-range Coulomb collisions involve the simultaneous interaction of many charged particles and are described by the more involved *Fokker–Planck* or *Landau collision operators.*

2.2 Relaxation model for molecular collisions

The formulation of specific models for the collision term $C_s(f_\alpha)$ we left aside in the previous section is a difficult task since the mathematical description of specific elementary processes are involved. This term in equations (2.9) or (2.10) accounts for the changes in time produced by molecular encounters in the velocity distribution function $f_\alpha(\boldsymbol{v}, \boldsymbol{r}, t)$. We will discuss the simple formulation of the *relaxation* or *Krook model*[7] which comes from the neutral gas particle collision term.

In the first place, the existence of a *local and stationary* distribution $f_\alpha^{eq}(\boldsymbol{r}, \boldsymbol{v})$ close to the local Maxwellian distribution (4.3) introduced in section 4.2 of volume 1 is assumed. The initial ($t = 0$) non-equilibrium distribution $f_{\alpha,o}(\boldsymbol{v}, \boldsymbol{r}, 0)$ evolves in time to f_α^{eq} at a rate,

$$\frac{\partial f_\alpha}{\partial t} = -\frac{(f_\alpha - f_\alpha^{eq})}{\tau_r} < 0$$

where $f_\alpha(\boldsymbol{v}, \boldsymbol{r}, t)$ is the distribution function for $t > 0$ and τ_r is a *characteristic relaxation time.* We obtain a differential equation for f_α as,

$$\frac{\partial f_\alpha}{\partial t} + \frac{f_\alpha}{\tau_r} = \frac{f_\alpha^{eq}}{\tau_r}$$

and the right term f_α^{eq}/τ_r is time-independent. Its complete solution is,

$$f_\alpha(\boldsymbol{v}, \boldsymbol{r}, t) = f_\alpha^{eq}(\boldsymbol{v}, \boldsymbol{r}) + C\, e^{-t/\tau_r}$$

and $C = (f_{\alpha,o} - f_\alpha^{eq})$ can be determined using the non-equilibrium distribution $f_{\alpha,o}(\boldsymbol{v}, \boldsymbol{r}, 0)$ for the initial instant giving,

$$f_\alpha(\boldsymbol{v}, \boldsymbol{r}, t) = f_\alpha^{eq}(\boldsymbol{v}, \boldsymbol{r}) + (f_{\alpha,o} - f_\alpha^{eq})\, e^{-t/\tau_r}$$

[7] Proposed by Bathnagar, Groos and Krook in reference [1], it is also called the BGK model.

The departure $(f_{\alpha,o} - f_{\alpha}^{eq})$ from the equilibrium distribution function decreases in time at a rate governed by the relaxation time τ_r and is also *spatially homogeneous* since it is independent of $\boldsymbol{r}$. Equivalently, the time-dependent amplitude $(f_{\alpha,o} - f_{\alpha}^{eq})$ of fluctuations from the equilibrium distribution function $f_{\alpha}^{eq}(\boldsymbol{r}, \boldsymbol{v})$ exponentially decays in time for small departures of the equilibrium state.

This simple *relaxation model* is practical and τ_c can be directly related with the dominant collision cross sections. In weakly ionized plasmas governed by short-range collisions of charged particles with neutral atoms we can introduce the relaxation times $\tau_{ea} \sim 1/\nu_{ea}$ and $\tau_{ia} \sim 1/\nu_{ia}$ for electron and ions. The rate of change of the ion and electron velocity distribution functions depend on the collision frequencies,

$$\nu_{ea} \sim n_a\, n_e\, \sigma_{ea}\, \bar{v}_i \text{ and also, } \nu_{ia} \sim n_a\, n_i\, \sigma_{ia}\, \bar{v}_i$$

introduced in section 3.4 of volume 1 that can be directly incorporated into this BGK model. Additionally, high neutral atom densities (proportional to the gas pressure p_a) reduce the relaxation times τ_r that govern the amplitude of fluctuations from the equilibrium plasma state. The short relaxation times facilitate a fast *collisional energy thermalization* among plasma species, as discussed in section 4.3 of volume 1.

As we shall see in chapter 3, this simple BGK model, can be employed to formulate the macroscopic momentum and energy exchange (close equilibrium state) between the electron and ion fluids in the hydrodynamic description of weakly ionized plasmas. However, except for systems where a specific collisional process dominates, this model is rather limited. Using a single time rate τ_r (equivalently, one collision cross section) the BGK model is a crude approximation that oversimplifies the relaxation phenomena of actual systems where multiple elementary processes can take place at the microscopic level, governed by different collision cross sections, etc.

2.3 The Boltzmann collision integral

The *Boltzmann collision integral* is a more refined approximation for the collision term $C_s(f_\alpha)$ in equation (2.10) than the BGK relaxation model but—as we shall see—it is still limited, since it is also based on simplifying assumptions. The Boltzmann collision integral model is applicable to sort-range binary collisions in dilute gases and plasmas where molecular or atomic forces decay much faster than the inter-particle distance. This situation takes place in weakly ionized plasmas and gases at low pressures. The spatial changes of velocity distributions $f_\alpha(\boldsymbol{v}, \boldsymbol{r}, t)$ are smooth and its temporal variations slow compared with characteristic collision time. That is, the model applies when the temporal scales of collisional processes faster than the evolution in time of the system. After its derivation we will examine with more detail the underlying physical approximations.

2.3.1 Qualitative derivation

In the following, we will consider a fixed *phase space* volume $d^3r\, d^3v_\alpha = d\boldsymbol{r}\, d\boldsymbol{v}_\alpha$ located at point $(\boldsymbol{r}, \boldsymbol{v}_\alpha)$ and the *geometry volume* $d\boldsymbol{r} = d^3r$ at point $\boldsymbol{r}$ will remain fixed and is occupied by target α-particles. Collisions with the incoming β-particles change the number of targets within the elementary volume $dv_\alpha = d^3v_\alpha$ placed about v_α in the *velocity space*.

The change in number of α-particles within the elementary phase space volume $(d^3r\ d^3v_\alpha)$ during the time interval δt can be expressed (see equation 2.1) as,

$$\delta N_\alpha = \delta N_\alpha^{\text{in}} - \delta N_\alpha^{\text{out}} = \left(\frac{\delta f_\alpha}{\delta t}\right)_{\alpha\beta} (d^3r\ d^3v_\alpha)\ \delta t$$

Here δN_α is the difference between the $\delta N_\alpha^{\text{in}}$ number of α-particles that come *into* the velocity volume interval d^3v_α located at point v_α *after* experiencing one collision event. The number $\delta N_\alpha^{\text{out}}$ are those initially within d^3v_α that are pulled out when colliding. Additionally, $(\delta f_\alpha/\delta t)_{\text{col}}$ represents one specific binary[8] (two-particle) collision term $C_{\alpha\beta}(f_\alpha, f_\beta)$ in the sum C_α of equation (2.10).

The term $\delta N_\alpha^{\text{out}}$ accounts for *direct collisions* that pull out the α-particle initially located within $(d^3r\ d^3v_\alpha)$ from the velocity range $(v_\alpha,\ v_\alpha + dv_\alpha)$ or equivalently, from the volume d^3v_α located at point v_α. The *inverse collisions* contribute to $\delta N_\alpha^{\text{in}}$ since they bring inside the volume d^3v_α placed at v_α the α-particles within d^3r others located outside, with initial velocities between v_α' and $v_\alpha' + dv_\alpha'$ before the encounter.

The number of incoming β-particles scattered by *one* α-target by *direct collisions* using equation (5.4) of volume 1 is,

$$\delta \dot{Q}_{\text{dir}} \times \delta t = [\,\sigma_{\text{dir}}(g, \chi)\ d\Omega\,] \times \Gamma_\beta \times \delta t$$

where $\Gamma_\beta = g\ f_\beta(v_\beta,\ \boldsymbol{r},\ t)\ d^3v_\beta$ is the flux (2.6) of β-particles, $\boldsymbol{g} = v_\beta - v_\alpha$ is the relative velocity, the collision cross section is $\sigma_{\text{dir}}(g, \chi)$ and $d\Omega$ the solid angle of figure 5.2 in volume 1. If we multiply by the number of α-particle targets within the d^3r geometry space volume,

$$\left[f_\alpha(v_\alpha,\ \boldsymbol{r},\ t)\ d^3v_\alpha\ d^3r\right] \times [\,\sigma_{\text{dir}}(g, \chi)\ d\Omega\,] \times g \times \left[f_\beta(v_\beta,\ \boldsymbol{r},\ t)\ d^3v_\beta\right] \times \delta t$$

we obtain the total number of encounters. Integrating for all possible velocities v_β and scattering angles $d\Omega$ we derive the contribution of direct collisions,

$$\delta N_\alpha^{\text{out}} = (\delta t\ d^3r) \times \int_{-\infty}^{+\infty} \int_\Omega g\ \sigma_{\text{dir}}(g, \chi)\ d\Omega\, f_\alpha\, f_\beta\ \ d^3v_\alpha\ d^3v_\beta \tag{2.14}$$

where we have introduced the notation $f_\alpha = f_\alpha(v_\alpha,\ \boldsymbol{r},\ t)$ and $f_\beta = f_\beta(v_\beta,\ \boldsymbol{r},\ t)$ for short.

A similar argumentation is used for the *inverse* collisions where again the β-particles scattered by one α-target is,

$$\delta \dot{Q}_{\text{inv}} \times \delta t = [\,\sigma_{\text{inv}}(g', \chi')\ d\Omega'\,] \times \Gamma_\beta' \times \delta t$$

[8] This specific approximation is crucial as collisions involving three or more particles are beyond this approach, valid only for low pressure (or *dilute*) gases.

where $\Gamma'_\beta = g' f_\beta(v'_\beta, \boldsymbol{r}, t)\, d^3v'_\beta$ is the flux of incoming particles and $\boldsymbol{g}' = v'_\beta - v'_\alpha$ their relative velocity. Multiplying by the α-targets inside d^3r we have,

$$\left[f_\alpha(v'_\alpha, \boldsymbol{r}, t)\, d^3v'_\alpha\, d^3r\right] \times [\sigma_{\rm inv}(g', \chi')\, d\Omega'] \times g' \times \left[f_\beta(v'_\beta, \boldsymbol{r}, t)\, d^3v'_\beta\right] \times \delta t$$

Integrating over all velocities v'_β and scattering angles $d\Omega'$ we obtain the contribution of direct collisions,

$$\delta N_\alpha^{\rm in} = (\delta t\, d^3r) \times \int_{-\infty}^{+\infty} \int_{\Omega'} g'\, \sigma_{\rm inv}(g', \chi')\, f'_\alpha f'_\beta\, d\Omega'\; d^3v'_\alpha\, d^3v'_\beta \tag{2.15}$$

where again $f'_\alpha = f'_\alpha(v'_\alpha, \boldsymbol{r}, t)$ and $f'_\beta = f'_\beta(v_\beta, \boldsymbol{r}, t)$. Finally, using equations (2.14) and (2.15) the collisional term can be calculated as,

$$C_{\alpha\beta}(f_\alpha, f_\beta) = \frac{\delta N_\alpha^{\rm in} - \delta N_\alpha^{\rm out}}{d^3v_\alpha\, d^3r\, \delta t} \tag{2.16}$$

that can be understood as a balance equation accounting for the number of incoming and outgoing particles from the elementary volume element of the phase space ($d^3r\, d^3v_\alpha$) caused by collisions. It is important to note that a specific model for the cross sections $\sigma_{\rm dir}(g, \chi)$ and $\sigma_{\rm dir}(g, \chi)$ is still required in equations (2.14) and (2.15). That is, the specific physical details of inter-particle interactions at the atomic and molecular level.

The above expression for $C_{\alpha\beta}(f_\alpha, f_\beta)$ takes a simple expression for *short range elastic collisions* where $g = g'$ since the vectors $\boldsymbol{g}$ and $\boldsymbol{g}'$ only differ in direction (see section 5.3 of volume 1) in this case. Additionally, $\sigma_{\rm dir}(g, \chi)\, d\Omega = \sigma_{\rm inv}(g', \chi')\, d\Omega'$ because the cross section is equal for forward and reverse collisions and assuming interparticle forces to have central symmetry is also invariant to an inversion of the coordinates[9]. Furthermore, using the center of mass $\boldsymbol{r}_{CM}$ we have,

$$dv_\alpha\, dv_\beta = d\boldsymbol{g}\, d\boldsymbol{r}_{cm} = d\boldsymbol{g}'\, d\boldsymbol{r}'_{cm} = dv'_\alpha\, dv'_\beta$$

With these simplifications we finally have,

$$\left(\frac{\delta f_\alpha}{\delta t}\right)_{\rm col} = \int_{-\infty}^{+\infty} \int_{\Omega} g\, \sigma_{\alpha\beta}(g, \chi)\, (f'_\alpha f'_\beta - f_\alpha f_\beta)\, d\Omega\, dv_\beta \tag{2.17}$$

It is important to note that *Boltzmann collision integral* couples the velocity distribution functions of two particle species through the collisional cross section of a specific elementary process. Then equation (2.10) is transformed into an integro-differential equation when collisions are taken into account as,

[9] The *direct collision* can be viewed replacing in figure 5.2 of volume 1 the target b by α and the incoming particle α is β. In the *reverse* encounter the incoming particle moves along the opposite direction that a point follows in the figure and symmetry gives equal solid angles for both processes.

$$\frac{\partial f_\alpha}{\partial t} + \nabla_{\mathrm{r}} \cdot (v_\alpha f_\alpha) + \nabla_{\mathrm{v}} \cdot (\boldsymbol{a} f_\alpha) = \int_{-\infty}^{+\infty} \int_\Omega g\, \sigma_{\alpha\beta}(g, \chi) \times \left(f'_\alpha f'_\beta - f_\alpha f_\beta\right) d\Omega\, dv_\beta \tag{2.18}$$

Each plasma species $\alpha = e, i, a$ is governed by a Boltzmann equation (2.10) with its specific collision integral (2.17) for each elementary process. The sum of all contributions gives,

$$C(f_\alpha) = \sum_{\beta \neq \alpha} C_{\alpha,\beta}(f_\alpha, f_\beta)$$

and therefore the velocity distribution functions of plasma species are coupled through collision integrals, that include the cross sections of collisional processes.

2.3.2 Approximations

The previous derivation of the Boltzmann equation (2.18) relies on the following relevant physical assumptions and/or approximations:

1. Only short-range *binary* collisions are taken into account and collective effects are ignored. This assumption is justified for low pressure gases or partially ionized low pressure plasmas where the collisions with neutral particles are dominant. This is not the case when the contribution of long-range Coulomb collisions are important.
2. The colliding particles are considered an isolated system in the above derivation. The effect of external forces on cross sections or in the two-particle collision parameters are neglected. The relative speed $g = g'$ is constant only in the absence of external forces and takes a cumbersome expression otherwise.
3. Within the phase space volume $(d^3r\; d^3v_\alpha)$ the velocity distribution functions are uniform and constant in time. The collision time interval δt is much shorter than the variation in time of the velocity distribution functions of particles and d^3r much smaller than the spatial variations of the distribution functions. In other words, the temporal scale of elementary processes are faster than the time evolution of the system.
4. The velocities and positions of particles before the encounter are independent (uncorrelated). The probability that an α-particle collides with the β-particle can be considered to be proportional to the product $f_\alpha \times f_\beta$ and correlations are neglected. This assumption is known as the *molecular chaos hypothesis.*

2.3.3 The Maxwell–Boltzmann distribution

The Boltzmann equation is valid for low pressure gases and plasmas where long-range Coulomb collisions can be neglected. This is usually the case of plasmas in space and technological applications where ionization degree is low and gas pressures are moderate. In these conditions the equilibrium Maxwell–Boltzmann velocity distribution (3.5) or (3.7) is recovered from the collision integral (2.17).

In the limit of thermodynamic equilibrium the velocity distribution function $f_\alpha(\boldsymbol{v}, \boldsymbol{r})$ is stationary and a sufficient condition is to satisfy $(\delta f_\alpha/\delta t)_{\rm col} = 0$ and then,

$$\int_{v_\beta}\int_\Omega g\,\sigma_{\alpha\beta}(g, \chi)\,(f'_\alpha f'_\beta - f_\alpha f_\beta)\,d\Omega\,dv_\beta = 0 \text{ then,} f'_\alpha f'_\beta - f_\alpha f_\beta = 0 \quad (2.19)$$

This functional equation has a *unique* solution that can be justified as follows. In the equilibrium state the distributions $f_\alpha(|v_\alpha|) = f_\alpha(v_\alpha^2)$ are *uniform* in space (independent of $\boldsymbol{r}$) and *isotropic* $f_\alpha(|v_\alpha|) = f_\alpha(v_\alpha^2)$ so we can write,

$$\ln f_\alpha(v_\alpha^2) + \ln f_\beta(v_\beta^2) = \ln f_\alpha(v'^2_\alpha) + \ln f_\beta(v'^2_\beta) \quad (2.20)$$

The unique solution possible to satisfy equations (2.19) and (2.20) and the conservation of energy in the collisions is,

$$f_\alpha(v_\alpha^2) = A\exp(B\,v_\alpha^2)$$

and the Maxwell–Boltzmann distribution (3.5) can be derived as in section 3.2 of volume 1 using the normalization condition for the A constant.

2.4 Commentaries and further reading

The books [2, 3] and [4] discuss in more detail the fundamentals of the plasma kinetic theory introduced in this chapter. The book [5] is an advanced text as is also the classical and comprehensive book [6]. Here we are mainly interested in partially ionized plasmas at low pressure where short-range collisions are dominant. The Boltzmann integral is in this case a valid physical description, and in chapter 21, sections 4.2 and 4.3 of reference [2] is introduced its application to ion and electron transport weakly ionized gases. This specific model is of interest in the physical description of low pressure electric discharges.

The Boltzmann equation and collision integral have played an important role connecting the deterministic laws of classical mechanics with the irreversible nature of non-equilibrium physical processes. The literature on the Boltzmann equation is huge and its mathematical properties are still a subject of research [7]. This chapter is only a brief overview to a challenging field with implications far beyond physics. Boltzmann's ideas have found applications in fields as diverse as econometric models [8], stock market trade [9], economic wealth [10] or social sciences [11].

The long-range Coulomb collisions require more involved models such as the Fokker–Planck and Landau collision operators. So far we are concerned with weakly ionized plasmas so they are outside the scope of the present work. These collision operators are discussed in chapter 21, sections 5.2 and 5.3 of the book [2] and also in chapter 3 of [12].

References

[1] Bathnagar P L, Gross E P and Krook M 1954 A model for collisional processes in gases. I. small amplitude processes in charged and neutral one-component systems *Phys. Rev.* **94** 511–25

[2] Bittencourt J A 2004 *Fundamentals of Plasma Physics* 3rd edn (New York: Springer)

[3] Huang K 1987 *Statistical Mechanics* 2nd edn (New York: Wiley)
[4] Nicholson D R 1983 *Introduction to Plasma Theory* 2nd edn (New York: Wiley)
[5] Woods L C 1993 *An Introduction to the Kinetic Theory of Gases and Magnetoplasmas* (Oxford: Oxford University Press)
[6] Chapman S and Cowling T G 1953 *Mathematical Theory of Non-uniform Gases* (Cambridge: Cambridge University Press)
[7] Harris S 2014 *An Introduction to the Theory of the Boltzmann equation* 2nd edn (New York: Dover)
[8] Golan A 2006 Information and entropy econometrics. A review and syntesis *Found. Trends Econometr* **2** 1–145
[9] Kanazawa K, Sueshige T, Takayasu H and Takayasu M 2018 Derivation of the Boltzmann equation for financial Brownian motion: Direct observation of collective motion of high frequency traders *Phys. Rev. Lett.* **120** 138301
[10] Boghosian B 2014 Kinetics of wealth and the Pareto law *Phys. Rev. E* **89** 042804
[11] Düring B, Markowich P, Pietschmann J-F and Wolfram M-T 2009 Boltzmann and Fokker–Plank equations modelling opinion formation in the presence of strong leaders *Proc. R. Soc. A* **465** 3687–708
[12] Fritzpatick R 2015 *Plasma Physics: An Introduction* (Boca Raton, FL: CRC Press)

IOP Publishing

An Introduction to Plasma Physics and its Space Applications, Volume 2
Basic equations and applications

Luis Conde

Chapter 3

Hydrodynamic description of plasmas

The hydrodynamic equations for a plasma can be derived from the Boltzmann equations for ions, electrons and neutral atoms by means of statistical averages. Each plasma species is described by an independent transport equation coupled to the others by macroscopic *friction* terms that account for their collisional interactions. This the complete hydrodynamic plasma equations are using physical assumptions as in the *cold* and *warm* plasma limits.

We have seen in chapter 2 that the mathematical formulation for collision operators $C(f_\alpha)$ for plasma species ($\alpha = e,\ i,\ a$) are cumbersome. The derivation of general expressions of non-equilibrium probability distribution functions $f_\alpha(\boldsymbol{v},\ \boldsymbol{r},\ t)$ as solutions of the kinetic equation (2.10) is a mathematical challenge. However, we can derive simpler and more intuitive fluid equations to describe the macroscopic transport of mass, energy and momentum in a plasma, which are valid under certain approximations.

The neutral gases can be seen as macroscopic continuous fluid when the dimensionless Knudsen number,

$$Kn = \frac{\lambda_c}{L_s} = \frac{k_B\, T}{\sqrt{2}\ \pi\ \sigma_c^2\ p_a\ L_s} \ll 1$$

is small, as shown in the scheme of figure 1.1. This approximation is valid, so far the characteristic dimension $L_s \gg \lambda_c$ of the physical system is much larger than the collisional mean free path λ_c for short-range molecular interactions[1]. The

[1] See equation volume 1, equation (2.2), box 2.2 and section 3.4.

macroscopic transport equations are simpler than kinetic theory and make use of a small number of physical parameters, such as the fluid velocity $\boldsymbol{u}(\boldsymbol{r}, t)$, temperature $T(\boldsymbol{r}, t)$ and the particle density $n(\boldsymbol{r}, t)$ or, equivalently, the pressure $p(\boldsymbol{r}, t)$. These magnitudes are statistical averages over a large number of particles[2] that change with position $\boldsymbol{r}$ and also evolve over time in the general case.

These assumptions can be extended to plasma theory and therefore, plasmas can be considered as *macroscopic continuous media* under certain conditions. The ions, electrons and neutral atoms can be regarded as three mutually inter-penetrating fluids, whose motions are coupled by both, short-range collisions and long-range electromagnetic interactions. In hydrodynamic models are introduced the densities $n_\alpha(\boldsymbol{r}, t)$, fluid velocities $\boldsymbol{u}_\alpha(\boldsymbol{r}, t)$ and local temperatures $T_\alpha(\boldsymbol{r}, t)$ of plasma species $(\alpha = a, e, i)$, similarly to transport equations of neutral fluids and gases.

In addition to the mean free path λ_c for the dominant collisional process, we need also to consider the Debye length scale[3] for the local electric field shielding. Furthermore, the number of charges $N_{ch} \sim n_\alpha \lambda_D^3 \gg 1$ needs to be large enough to account for the Debye shielding.

The hydrodynamic approximation for a plasma is appealing when a large number of charged particles are into a small cube with lateral size $l_c \gg \lambda_c$ much smaller than the relevant macroscopic dimension of the system $L_s \gg l_c \gtrsim \lambda_D$, but longer than the Debye length. We have $n_\alpha l_c^3 \gg 1$ and also $n_\alpha \lambda_D^3 \gg 1$, in this case the averaged physical magnitudes such as the electric charge density $\rho_c(\boldsymbol{r}, t)$ or current density $\boldsymbol{J}_c(\boldsymbol{r}, t)$ are statistically valid concepts.

Fluid models basically apply to weakly coupled LTE plasmas where particle collisions along $l_c \ll L_s$ relax the random fluctuations of energy $\delta E_\alpha \sim \delta T_\alpha \ll T_\alpha$ and particle densities $\delta n_\alpha \ll n_\alpha$ over a time scale $\tau \ll L_s/|\boldsymbol{u}_\alpha|$ much shorter than the macroscopic plasma motion. Furthermore, the Debye shielding over distances $\lambda_D \ll l_c$ attenuates the local fluctuations of the electric field within the small l_c^3 volume. PLE plasmas can be also described using hydrodynamic transport equations for distances L_c much longer than the scale length l_c introduced by the dominant collisional process.

Box 3.1 shows that a number of plasmas in Nature and in the laboratory where local fluctuations decay over small distances l_c compared with the characteristic transport length scale L_s satisfy these conditions. The opposite limit where $\lambda_c \gtrsim L_s$ requires a kinetic approach using the Boltzmann or Vlasov equations.

The fluid approximation assumes smooth changes of the non-equilibrium velocity distribution function $f_\alpha(\boldsymbol{v}, \boldsymbol{r}, t)$ along the length $L_s \gg l_c$ and time $T_s \gg l_c/|\boldsymbol{u}_\alpha|$ scales involved in the macroscopic plasma transport. In these conditions the *hydrodynamic plasma transport equations* governing the temporal and spatial evolution of $n_\alpha(\boldsymbol{r}, t)$, $\boldsymbol{u}_\alpha(\boldsymbol{r}, t)$ and $T_\alpha(\boldsymbol{r}, t)$, can be directly derived from the Boltzmann equation (2.10)

[2] The equivalent time-independent averages were introduced in section 3.3 of volume 1.

[3] Strictly speaking, the Debye length expression (4.13) in volume 1 is derived for Maxwellian plasmas. This is not the general case, so in its place we can consider λ_D as an *effective length scale* for electric field shielding along which plasma quasineutrality is preserved.

Box 3.1. The fluid approximation

A high pressure argon arc has a typical kinetic temperature $T \sim 0.5$ eV and plasma density $n_e \sim 10^{15}$ cm^{-3} for the gas pressure of $p_a \sim 50$ mbar in figures 1.2 and 4.1 of volume 1. This partially ionized ($\alpha_g \sim 0.08\%$) LTE plasma has $\lambda_D \sim 1.7 \times 10^{-5}$ cm and in figure 6.3(a) of volume 1 the cross section is of the order of $\sigma_{el} \sim 10^{-15}$ cm^2. These values give the mean free path $\lambda_c \sim 8.3 \times 10^{-4}$ cm for the short-range elastic collisions.

Therefore, for the length scale $l_c \sim 10^{-2}$ cm $\gtrsim \lambda_D \sim 10^{-3}$ cm the number of electrons inside l_c^3 is,

$$N_e = l_c^3 \times n_e = (10^{-2})^3 \times 10^{15} = 10^9 \text{ charged particles.}$$

Therefore, the fluid approximation $L_s \gg l_c \gtrsim \lambda_D$ remains valid when the macroscopic dimensions of the physical system are $L_s \gtrapprox 0.1$ cm, which is small for the length scales involved in mass transport in arc plasmas.

The density of charged particles in figure 1.7 of volume 1 for the ionospheric F-peak located at 200 km of altitude is $n \sim 2.5 \times 10^5$ particles by cubic centimeter with $T_e \sim 0.05$ eV which gives $\lambda_D \sim 0.3$ cm (see table 4.1, volume 1). The average of the ion mass at this height is approximated by 24 amu[a] [1] and in this fully ionized plasma, the long-range Coulomb collision frequency of an electron by ions can be estimated using equation (5.22) of volume 1,

$$\nu_{ei} = \frac{e^4\, n_e \ln \Lambda_{cl}}{6\sqrt{2}\, \epsilon_o^2\, \pi^{3/2}\, m_e^{1/2}\, (k_B T)^{3/2}} \quad \text{and} \quad \nu_{ie} \sim \frac{m_e}{m_i}\nu_{ei} \ll \nu_{ei}$$

for the deflection of ions by electrons. Using for the Coulomb logarithm $\ln \lambda_c = 15$ we obtain $\nu_{ie} \sim 0.46$ s^{-1} and $\nu_{ei} \sim 10^{-3}$ s^{-1}, hence $\lambda_{ie} = c_{is}/\nu_{ie} \sim 4$ km, where c_{is} is the ion sound speed.

Hence, $l_c \sim 4 \times 10^5$ cm $\gg \lambda_D \sim 0.3$ cm and the number of electrons inside l_c^3 is now,

$$N_e = l_c^3 \times n_e = (10^5)^3 \times 2.5 \times 10^5 = 2.5 \times 10^{20} \text{ charged particles.}$$

In this case, the macroscopic length scale $L_s \gg l_c \gtrsim \lambda_D$ is much longer than the collision mean free path for ions, in the order of tens of kilometers. However, this value is small compared with the the characteristic lengths involved in the ionospheric mass transport.

without using an explicit expression for the distribution function. However, the specific details of particle collisions considered in the operator $C(f_\alpha)$ are required to derive the macroscopic transport coefficients, such as thermal conductivity, viscosity, etc.

[a] 1 amu = 1.66×10^{-27} kg.

3.1 The moments of Boltzmann equation

The *macroscopic transport equations* that govern the temporal and spatial variations of density, fluid velocity and kinetic temperature can be derived by the mathematical procedure of *taking moments* of the Boltzmann equation (2.10). The *moment* $\boldsymbol{M}^{(k)}(\boldsymbol{r}, t)$ of order k of the distribution function $f(\boldsymbol{v}, \boldsymbol{r}, t)$ can be formally written as,

$$\boldsymbol{M}^{(k)}(\boldsymbol{r}, t) = \int_{-\infty}^{+\infty} \underbrace{(\boldsymbol{v} \otimes \boldsymbol{v} \otimes \ldots \otimes \boldsymbol{v})}_{\text{k times}} f(\boldsymbol{v}, \boldsymbol{r}, t)\, d\boldsymbol{v} \tag{3.1}$$

where $\boldsymbol{a} \otimes \boldsymbol{b}$ denotes the tensor product of vectors $\boldsymbol{a}$ and $\boldsymbol{b}$. The infinite set of moments $\boldsymbol{M}^{(k)}(\boldsymbol{r}, t)$ for $k = 0, 1, 2, \ldots$ characterizes the function $f(\boldsymbol{v}, \boldsymbol{r}, t)$ when this distribution is smooth enough.

These *moments* are the mathematical generalization of the averages in the phase space introduced in chapter 3 of volume 1 and chapter 2 of this volume. For the particle species α these averages in the $(\boldsymbol{r}, \boldsymbol{v})$ space can be generalized for the time dependent $H_\alpha(\boldsymbol{v}, \boldsymbol{r}, t)$ function as,

$$\langle H_\alpha(\boldsymbol{v}, \boldsymbol{r}, t)\rangle = \frac{1}{n_\alpha(\boldsymbol{r}, t)} \int_{-\infty}^{+\infty} H_\alpha(\boldsymbol{v}, \boldsymbol{r}, t) f_\alpha(\boldsymbol{v}, \boldsymbol{r}, t)\, d\boldsymbol{v} \tag{3.2}$$

where again the index $(\alpha = a, e, i)$ indicates neutral atoms, electrons or ions.

The connection of these abstract mathematical expressions with magnitudes of physical interest can be grasped with the simplest choice $H_\alpha(\boldsymbol{v}, \boldsymbol{r}, t) = 1$ corresponding to $k = 0$ in equation (3.2). We obtain,

$$1 = \frac{1}{n_\alpha(\boldsymbol{r}, t)} \int_{-\infty}^{+\infty} f_\alpha(\boldsymbol{v}, \boldsymbol{r}, t)\, d\boldsymbol{v} \quad \text{or,} \quad n_\alpha(\boldsymbol{r}, t) = \int_{-\infty}^{+\infty} f_\alpha(\boldsymbol{v}, \boldsymbol{r}, t)\, d\boldsymbol{v} \tag{3.3}$$

which is related with the number density $n_\alpha(\boldsymbol{r}, t)$ introduced in equation (2.4). Equivalently,

$$1 = \int_{-\infty}^{+\infty} P_\alpha(\boldsymbol{v}, \boldsymbol{r}, t)\, d\boldsymbol{v}$$

is the normalization of the probability distribution function $P_\alpha = f_\alpha(\boldsymbol{v}, \boldsymbol{r}, t)/n_\alpha(\boldsymbol{r}, t)$ introduced in equation (2.2). Setting $k = 1$ in equation (3.1) or $H_\alpha(\boldsymbol{v}, \boldsymbol{r}, t) = \boldsymbol{v}$ in equation (3.2) we have,

$$\langle \boldsymbol{v}\rangle = \frac{1}{n_\alpha(\boldsymbol{r}, t)} \int_{-\infty}^{+\infty} \boldsymbol{v} f_\alpha(\boldsymbol{v}, \boldsymbol{r}, t)\, d\boldsymbol{v} \tag{3.4}$$

Then $\langle \boldsymbol{v}\rangle = \boldsymbol{u}_\alpha(\boldsymbol{r}, t)$ is the average macroscopic fluid velocity introduced in equation (2.5). With $H_\alpha(\boldsymbol{v}, \boldsymbol{r}, t) = m_\alpha\, (\boldsymbol{v} \otimes \boldsymbol{v})$ or $k = 2$ in equation (3.1) we have,

$$m_\alpha\, \langle \boldsymbol{v} \otimes \boldsymbol{v}\rangle = \frac{1}{n_\alpha(\boldsymbol{r}, t)} \int_{-\infty}^{+\infty} m_\alpha\, (\boldsymbol{v} \otimes \boldsymbol{v}) f_\alpha(\boldsymbol{v}, \boldsymbol{r}, t)\, d\boldsymbol{v} \tag{3.5}$$

and the components of the *momentum flow tensor* $\boldsymbol{S}_\alpha = m_\alpha\, n_\alpha(\boldsymbol{r},\, t)\, \langle \boldsymbol{v} \otimes \boldsymbol{v}\rangle$ are,

$$\begin{pmatrix} v_x^2 & v_xv_y & v_xv_z \\ v_yv_x & v_y^2 & v_yv_z \\ v_zv_x & v_zv_y & v_z^2 \end{pmatrix} \tag{3.6}$$

where it is clear that $S_{ij} = S_{ji}$ so the matrix is symmetric.

The scalar function $H_\alpha(\boldsymbol{v},\, \boldsymbol{r},\, t) = m_\alpha\, (\boldsymbol{v} \cdot \boldsymbol{v})/2 = m_\alpha v^2/2$ gives the average kinetic energy per particle,

$$\left\langle \frac{m_\alpha\, v^2}{2} \right\rangle = \frac{1}{n_\alpha(\boldsymbol{r},\, t)} \int_{-\infty}^{\infty} \left(\frac{m_\alpha\, v^2}{2} \right) f_\alpha(\boldsymbol{v},\, \boldsymbol{r},\, t)\, d\boldsymbol{v} \tag{3.7}$$

whereas $H_\alpha(\boldsymbol{v},\, \boldsymbol{r},\, t) = (m_\alpha v^2/2)\, \boldsymbol{v}$ leads to,

$$\left\langle \left(\frac{m_\alpha\, v^2}{2} \right) \boldsymbol{v} \right\rangle = \frac{1}{n_\alpha(\boldsymbol{r},\, t)} \int_{-\infty}^{+\infty} \left(\frac{m_\alpha\, v^2}{2} \right) \boldsymbol{v}\, f_\alpha(\boldsymbol{v},\, \boldsymbol{r},\, t)\, d\boldsymbol{v} \tag{3.8}$$

which represents the average *flux of kinetic energy* transported per particle.

The physical description of macroscopic fluids is therefore related to the first moments of equation (3.1), that are connected to the averages in phase space over the velocity distribution function. At this point we face for the first time the problem of the *closure* of transport equations as the rigorous mathematical calculation requires the determination of an *infinite number of moments* given by equation (3.1). The time evolution of n_α depends on the velocity $\boldsymbol{u}_\alpha$, which in turn is a function of $\boldsymbol{S}_\alpha$ and so on; each one depends on the following higher order moment.

However, the hydrodynamic transport equations make use only of few averages (3.3)–(3.16), which are related to the magnitudes of physical significance. As we shall see, the derivation of transport equations requires a *truncation scheme* or further physical assumptions to cope with the reduced number of moments (3.1) of physical interest.

Before deriving the hydrodynamic equations of a plasma we will introduce few additional concepts. First, we consider the infinitesimal volume element P of figure 3.1 centered at point $\boldsymbol{r}$ of small size l_c but big enough to contain a large number $n_\alpha\, l_c^3 \gg 1$ of particles inside. Its center of mass moves with respect to the laboratory frame with average velocity $\boldsymbol{u}_\alpha(\boldsymbol{r},\, t)$ (2.5) or (3.4) along the dotted path line.

The velocity of a specific particle $\boldsymbol{v} = \boldsymbol{c} + \boldsymbol{u}_\alpha$ can be expressed as the sum of $\boldsymbol{u}_\alpha$ and a *randomly fluctuating component* $\boldsymbol{c}$ called *peculiar velocity* relative to a frame that moves with P with the $\boldsymbol{u}_\alpha$ velocity. Since $\boldsymbol{c}$ characterized the random motion of particles with respect to P its average[4] (3.2),

[4] We have $d\boldsymbol{v} = d\boldsymbol{c}$ and we also write $d\boldsymbol{v} = d^3v = d\boldsymbol{c} = d^3c$. Equivalently $\langle \boldsymbol{v}\rangle = \langle \boldsymbol{c}\rangle + \langle \boldsymbol{v}\rangle$ gives $\boldsymbol{u}_\alpha = \langle \boldsymbol{c}\rangle + \boldsymbol{u}_\alpha$ and then $\langle \boldsymbol{c}\rangle = 0$.

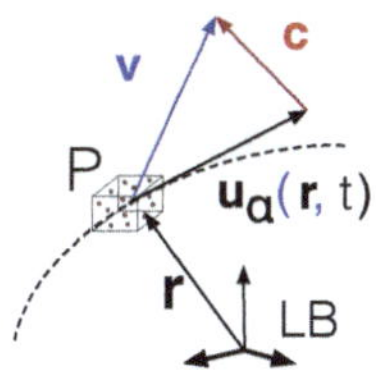

Figure 3.1. The small volume P moves with respect to the laboratory frame.

$$\langle \boldsymbol{c} \rangle = \frac{1}{n_\alpha(\boldsymbol{r}, t)} \int_{-\infty}^{+\infty} \boldsymbol{c}\, f_\alpha(\boldsymbol{v}, \boldsymbol{r}, t)\, d\boldsymbol{v} = 0$$

is null. In contrast, $\langle c^2 \rangle$ is related to the average kinetic energy with respect to the center of mass of P and,

$$\left\langle \frac{m_\alpha}{2} c^2 \right\rangle = \frac{1}{n_\alpha(\boldsymbol{r}, t)} \int_{-\infty}^{+\infty} \frac{m_\alpha}{2} c^2 f_\alpha(\boldsymbol{v}, \boldsymbol{r}, t)\, d\boldsymbol{v} = \frac{m_\alpha}{2} \langle c^2 \rangle > 0 \tag{3.9}$$

is positive since particles are always in thermal motion. For the average velocity (3.4) we can write,

$$\langle v^2 \rangle = \langle (\boldsymbol{u}_\alpha + \boldsymbol{c})^2 \rangle = \langle \boldsymbol{u}_\alpha^2 + 2\,(\boldsymbol{u}_\alpha \cdot \boldsymbol{c}) + \boldsymbol{c}^2 \rangle = u_\alpha^2 + \langle c^2 \rangle$$

Using equation (3.7) the average kinetic energy is,

$$E_\alpha(\boldsymbol{r}, t) = \int_{-\infty}^{\infty} \left(\frac{m_\alpha\, v^2}{2} \right) f_\alpha(\boldsymbol{v}, \boldsymbol{r}, t)\, d\boldsymbol{v} = n_\alpha(\boldsymbol{r}, t) \left\langle \frac{m_\alpha\, v^2}{2} \right\rangle$$

and we have,

$$E_\alpha = n_\alpha \left\langle \frac{m_\alpha}{2} v^2 \right\rangle = n_\alpha\, \frac{m_\alpha}{2}\, (u_\alpha^2 + \langle c^2 \rangle) = m_\alpha\, n_\alpha\, \frac{u_\alpha^2}{2} + e_{i\alpha} \tag{3.10}$$

The first term represents the average kinetic energy of particle by unit of volume with respect to the laboratory frame. The second adding $e_{i\alpha} = m_\alpha\, n_\alpha\, \langle c^2 \rangle / 2$ in equation (3.10) involves the random *peculiar velocity* and is the kinetic energy per volume unit in a frame moving with the small volume P and is a function of the kinetic temperature. Equation (3.10) states that increments in the energy of the small volume P, can raise its internal energy $e_{i\alpha}$ (temperature T_α), its velocity $\boldsymbol{u}_\alpha$ with respect to the laboratory frame or both.

When the particles in P can be considered in a *local equilibrium* $e_{i\alpha} \sim T_\alpha$ and for a local Maxwellian distribution (4.3) of volume 1, we have the internal energy[5],

$$\left\langle \frac{m_\alpha\, c^2}{2} \right\rangle = \frac{3}{2}\, k_B\, T_\alpha(\boldsymbol{r}, t) \quad \text{and,} \quad e_{i\alpha} = \frac{3}{2}\, n_\alpha(\boldsymbol{r}, t)\, k_B\, T_\alpha(\boldsymbol{r}, t) \tag{3.11}$$

[5] We make use of the internal energy of a monoatomic ideal, for molecules the factor 3/2 is replaced by $1/(\gamma - 1)$ where γ is the adiabatic index.

Equivalently, this medium is an LTE plasma where the equation,

$$p_\alpha(\boldsymbol{r},\, t) = n_\alpha(\boldsymbol{r},\, t)\, k_B T_\alpha(\boldsymbol{r},\, t)$$

is valid as far as short and long-range collisions relax fluctuations back to the stationary equilibrium state much faster than the $l_s/|\boldsymbol{u}_\alpha|$ slow time scale. This principle also applies to PLE plasmas under appropriate length and time scales.

Introducing $\boldsymbol{v} = \boldsymbol{u}_\alpha + \boldsymbol{c}$ in the expression for $\boldsymbol{S}_\alpha$, the averages in the matrix (3.5) are of the form,

$$\langle (u_{\alpha i} + c_i)(u_{\alpha j} + c_j)\rangle = \langle u_{\alpha i}\, u_{\alpha j}\rangle + \langle c_i\, c_j\rangle + u_{\alpha i}\, \langle c_j\rangle + u_{\alpha j}\, \langle c_i\rangle$$

Since $\langle c_i\rangle = 0$ are null for the components of the random velocity we have,

$$\boldsymbol{S}_\alpha = m_\alpha\, n_\alpha\, (\boldsymbol{u}_\alpha \otimes \boldsymbol{u}_\alpha) + m_\alpha\, n_\alpha \langle \boldsymbol{c} \otimes \boldsymbol{c}\rangle$$

and $\boldsymbol{S}_\alpha$ is called *momentum flow tensor* for the α-particles. The second is the *kinetic pressure tensor* $\boldsymbol{P}_\alpha$ term,

$$\boldsymbol{P}_\alpha = m_\alpha\, n_\alpha \langle \boldsymbol{c} \otimes \boldsymbol{c}\rangle = \boldsymbol{D}_\alpha + \boldsymbol{\Pi}_\alpha$$

and its components are similar to the matrix (3.6) replacing v_{ij} by c_{ij}. It can be cast as the sum of the $\boldsymbol{D}_\alpha$ and $\boldsymbol{\Pi}_\alpha$ tensors where,

$$\boldsymbol{D}_\alpha = m_\alpha\, n_\alpha \begin{pmatrix} \langle c_y^2\rangle & 0 & 0 \\ 0 & \langle c_y^2\rangle & 0 \\ 0 & 0 & \langle c_y^2\rangle \end{pmatrix} \tag{3.12}$$

Under LTE conditions the distribution of random peculiar velocities $\boldsymbol{c}$ is *isotropic* so,

$$\langle c_x^2\rangle = \langle c_y^2\rangle = \langle c_z^2\rangle = \frac{\langle c^2\rangle}{3}$$

and the tensor $\boldsymbol{D}_\alpha$ becomes proportional to a diagonal matrix. Using the ideal gas approximation (3.11) we have $\langle c^2\rangle = 3\, k_B\, T_\alpha/m_\alpha$ and for one component,

$$m_\alpha\, n_\alpha\, \langle c_x^2\rangle = m_\alpha\, n_\alpha\, \frac{\langle c^2\rangle}{3} = n_\alpha k_B\, T_\alpha$$

The pressure $p_\alpha(\boldsymbol{r},\, t)$ in equation (3.12),

$$p_\alpha(\boldsymbol{r},\, t) = \frac{m_\alpha}{3} \int_{-\infty}^{+\infty} c^2 f_\alpha(\boldsymbol{v},\, \boldsymbol{r},\, t)\, d\boldsymbol{v} = \frac{m_\alpha n_\alpha}{3} \langle c^2\rangle \tag{3.13}$$

represents the scalar *partial pressure* due to the particles of type α. The second tensor $\boldsymbol{\Pi}_\alpha$ is related with viscous stresses as in conventional fluid theory and its components are,

$$\boldsymbol{\Pi}_\alpha = m_\alpha\, n_\alpha \begin{pmatrix} 0 & \langle c_x\, c_y\rangle & \langle c_x\, c_z\rangle \\ \langle c_y\, c_x\rangle & 0 & \langle c_y\, c_z\rangle \\ \langle c_z\, c_x\rangle & \langle c_z\, c_y\rangle & 0 \end{pmatrix} \tag{3.14}$$

where $\langle c_i\, c_j\rangle = \langle c_j\, c_i\rangle$. Finally, the *momentum flow tensor* reads,

$$\boldsymbol{S}_\alpha = m_\alpha\, n_\alpha\, \langle \boldsymbol{v} \otimes \boldsymbol{v}\rangle = m_\alpha\, n_\alpha\, (\boldsymbol{u}_\alpha \otimes \boldsymbol{u}_\alpha) + \left(p_\alpha\, \boldsymbol{I}\right) + \boldsymbol{\Pi}_\alpha \tag{3.15}$$

where $\boldsymbol{I}$ is the unity matrix and the ideal gas law relation was also employed, equivalent to a local Maxwellian velocity distribution function for the LTE equilibrium of α-particles.

The non-diagonal terms can be neglected as the averages $\langle c_i\, c_j\rangle$ in tensor $\boldsymbol{\Pi}_\alpha$ are small under LTE conditions and $\boldsymbol{P}_\alpha \simeq (p_\alpha(\boldsymbol{r},\, t)\, \boldsymbol{I})$ for isotropic velocity distributions of peculiar (random) velocities.

The *energy density flux* vector $\boldsymbol{K}_\alpha(\boldsymbol{r},\, t)$ is introduced from equation (3.8) as,

$$\boldsymbol{K}_\alpha(\boldsymbol{r},\, t) = \int_{-\infty}^{+\infty} \left(\frac{m_\alpha\, v^2}{2}\right) \boldsymbol{v}\, f_\alpha(\boldsymbol{v},\, \boldsymbol{r},\, t)\, d\boldsymbol{v} = \frac{m_\alpha n_\alpha}{2}\, \langle v^2\, \boldsymbol{v}\rangle \tag{3.16}$$

and using $\boldsymbol{v} = \boldsymbol{c} + \boldsymbol{u}_\alpha$ in the average, the non-null terms are,

$$\langle v^2\, \boldsymbol{v}\rangle = (u_\alpha^2 + \langle c^2\rangle)\, \boldsymbol{u}_\alpha + \langle c^2\, \boldsymbol{c}\rangle + 2\, \langle\, (\boldsymbol{c} \cdot \boldsymbol{u}_\alpha)\, \boldsymbol{c}\rangle$$

The first addition at the right side is proportional to the transport $E_\alpha\, \boldsymbol{u}_\alpha$ of energy introduced in equation (3.10). The second term is the *heat flux vector* $\boldsymbol{q}_\alpha$ for the α species which is defined as,

$$\boldsymbol{q}_\alpha = \int_{-\infty}^{+\infty} \left(\frac{m_\alpha c^2}{2}\right) \boldsymbol{c}\, f_\alpha(\boldsymbol{v},\, \boldsymbol{r},\, t)\, d\boldsymbol{v} = \frac{m_\alpha\, n_\alpha}{2}\, \langle c^2\, \boldsymbol{c}\rangle \tag{3.17}$$

The vector $\boldsymbol{A} = \langle(\boldsymbol{c} \cdot \boldsymbol{u}_\alpha)\, \boldsymbol{c}\rangle$ in the last term is,

$$\boldsymbol{A} = \left\langle \left(\sum_j c_j\, u_{\alpha j}\right) \boldsymbol{c} \right\rangle$$

and for its k component of $\boldsymbol{c}$ we have,

$$A_k = \left\langle \left(\sum_j c_j\, u_{\alpha j}\right) c_k \right\rangle = \sum_j \langle c_j\, c_k\rangle\, u_{\alpha j}$$

this is equivalent to the contraction operation of the symmetric second order *momentum flux tensor* $m_\alpha\, n_\alpha \langle \boldsymbol{c} \otimes \boldsymbol{c}\rangle$ sum of equations (3.12) and (3.12) with the velocity $\boldsymbol{u}_\alpha$ we denote[6] as,

[6] Here the operation $\boldsymbol{A}$: $\boldsymbol{B}$ denotes the contraction of second order tensor $\boldsymbol{A}$ with vector $\boldsymbol{B}$.

$$A = \langle (\boldsymbol{c} \cdot \boldsymbol{u}_\alpha)\, \boldsymbol{c} \rangle = (P_\alpha \boldsymbol{I} + \boldsymbol{\Pi}_\alpha) : \boldsymbol{u}_\alpha$$

and p_α $(\mathbf{1}: \boldsymbol{u}_\alpha) = (p_\alpha\, \boldsymbol{u}_\alpha)$. Finally, the *energy density flux vector* becomes,

$$\boldsymbol{K}_\alpha(\boldsymbol{r}, t) = E_\alpha\, \boldsymbol{u}_\alpha + \boldsymbol{q}_\alpha + (p_\alpha\, \boldsymbol{u}_\alpha) + (\boldsymbol{\Pi}_\alpha : \boldsymbol{u}_\alpha) \tag{3.18}$$

The following derivation of the macroscopic transport equations needs equations (3.15), (3.17) and (3.18).

3.1.1 The equation of continuity

The equation for the conservation of the number (or mass density) of particles for the α species is obtained integrating equation (2.10) over the velocities v in the phase space. We have,

$$\int_{-\infty}^{+\infty} \frac{\partial f_\alpha}{\partial t}\, dv + \int_{-\infty}^{+\infty} \nabla_{\mathrm{r}} \cdot (v\, f_\alpha)\, dv + \int_{-\infty}^{+\infty} \nabla_{\mathrm{v}} \cdot \left(\frac{\boldsymbol{F}_{e\alpha}}{m_\alpha} f_\alpha \right) dv = \int_{-\infty}^{+\infty} C_s(f_\alpha)\, dv$$

where $\boldsymbol{F}_{e\alpha} = q_\alpha\, [\boldsymbol{E} + (v \times \boldsymbol{B})]$ is the Lorentz force acting on the α-particle and $\boldsymbol{E}(\boldsymbol{r}, t)$ and $\boldsymbol{B}(\boldsymbol{r}, t)$ the electromagnetic fields. Integrals in two first terms on the right can be directly permuted with the operators $\partial/\partial t$ and ∇_{r} using the divergence theorem for the third adding we have,

$$\frac{\partial}{\partial t} \int_{-\infty}^{+\infty} f_\alpha\, d\boldsymbol{v} + \nabla_{\mathbf{r}} \cdot \int_{-\infty}^{+\infty} (\boldsymbol{v}\, f_\alpha)\, d\boldsymbol{v} + \underbrace{\int_{s(v)} \left(\frac{\boldsymbol{F}_{e\alpha}}{m_\alpha} f_\alpha \right) \cdot d\boldsymbol{S}}_{0} = \int_{-\infty}^{+\infty} C_s(f_\alpha)\, d\boldsymbol{v}$$

The third integral is null since the velocity distribution function decreases as the velocity v grows increasing the surface $S(v)$ in the velocity space and equations (2.4) and (2.5) give,

$$n_\alpha(\boldsymbol{r}, t) = \int_{-\infty}^{+\infty} f_\alpha(v, \boldsymbol{r}, t)\, dv \quad \text{and} \quad n_\alpha\, \boldsymbol{u}_\alpha = \int_{-\infty}^{+\infty} v\, f_\alpha(v, \boldsymbol{r}, t)\, dv$$

Finally we obtain for the macroscopic density of α species,

$$\frac{\partial n_\alpha}{\partial t} + \nabla_{\mathrm{r}} \cdot (n_\alpha \boldsymbol{u}_\alpha) = \int_{-\infty}^{+\infty} C_\alpha(f_\alpha)\, dv$$

where the right term is the moment of the collisional operator. In section 2.3 we introduced equation (2.16) as the number of α-particles added or removed in time δt from the phase space volume $d^3v\, d^3r$ located at point $(\boldsymbol{r}, t)$ due to collisions with the β species. We can write,

$$C_\alpha = \frac{\delta N_\alpha^{\mathrm{in}} - \delta N_\alpha^{\mathrm{out}}}{d^3v\, d^3r\, \delta t}$$

for the creation and recombination of α-particles. Then,

$$\int_{-\infty}^{+\infty} C_\alpha(f_\alpha)\, dv = \frac{\delta N_\alpha^{\mathrm{in}} - \delta N_\alpha^{\mathrm{out}}}{d^3r\, \delta t} = \frac{\delta n_\alpha^{\mathrm{in}}}{\delta t} - \frac{\delta n_\alpha^{\mathrm{out}}}{\delta t} = S_\alpha - L_\alpha$$

where S_α and L_α represent the gain and loss rates per unit volume of geometry space. In the case of ions and electrons, these are the ionization $S_i = \langle k_I \rangle\, n_a\, n_e$ and recombination $L_i = \beta_R\, n_e\, n_i$ collision frequencies, equations (6.4) and (6.5) in volume 1. In the plasma equilibrium, the α-particles are lost and created at equal rates and $S_\alpha = L_\alpha$ expresses the balance of the corresponding forward and backward elementary processes.

Finally, the fluid particle continuity equation for α-particles reads,

$$\frac{\partial n_\alpha}{\partial t} + \nabla_{\mathrm{r}} \cdot (n_\alpha\, \boldsymbol{u}_\alpha) = S_\alpha - L_\alpha \tag{3.19}$$

The continuity equation expresses the conditions for the conservation of ($\alpha = a, e, i$) particles in each fluid element, similarly to conventional hydrodynamics. In this equation the different plasma species are coupled through the particle source/sink terms on the right hand. In this derivation no explicit expression for the velocity distribution function $f_\alpha(\boldsymbol{v}, \boldsymbol{r}, t)$ has been used. The statistical averages (2.4) and (2.5) only require the assumption of a bounded smooth $f_\alpha(\boldsymbol{v}_\alpha, \boldsymbol{r}, t)$ velocity distribution function.

In contrast, the calculation of S_α and L_α in equations (6.4) and (6.5) of volume 1 need an *explicit expression* for $f_\alpha(\boldsymbol{v}_\alpha, \boldsymbol{r}, t)$ and physical models for elementary processes such as the ionizing, recombination collisions, etc, at the atomic and molecular level. Therefore, as is indicated in the scheme of figure 1.1, the *mathematical closure* of plasma fluid equations will require *additional information* and/or further approximations.

3.1.2 The momentum transport equation

The next moment of the Boltzmann equation with physical meaning is obtained by multiplying the equation (2.10) by $m_\alpha \boldsymbol{v}$ and integrating over the velocities $\boldsymbol{v}$ as,

$$\begin{aligned}\int_{-\infty}^{+\infty} (m_\alpha\, \boldsymbol{v})\, \frac{\partial f_\alpha}{\partial t}\, d\boldsymbol{v} &+ \int_{-\infty}^{+\infty} (m_\alpha\, \boldsymbol{v})\, \nabla_{\mathrm{r}} \cdot (\boldsymbol{v}\, f_\alpha)\, d\boldsymbol{v} \\ &+ \int_{-\infty}^{+\infty} (m_\alpha\, \boldsymbol{v})\, \nabla_{\mathrm{v}} \cdot (\boldsymbol{F}_{e\alpha}\, f_\alpha)\, d\boldsymbol{v} = \\ &\qquad \int_{-\infty}^{+\infty} (m_\alpha\, \boldsymbol{v})\, C_\alpha(f_\alpha)\, d\boldsymbol{v}\end{aligned} \tag{3.20}$$

where $\boldsymbol{F}_{e\alpha} = q_\alpha\, [\boldsymbol{E} + (\boldsymbol{v} \times \boldsymbol{B})]$ is the Lorentz force acting on the α-particle. The right term involves the collision operator $C_s(f_\alpha)$ and can be simply expressed as the vector,

$$\boldsymbol{R}_\alpha = \int_{-\infty}^{+\infty} (m_\alpha\, \boldsymbol{v})\, C_\alpha(f_\alpha)\, d\boldsymbol{v} \tag{3.21}$$

that will be discussed later in section 3.1.3.

The first term in equation (3.20) is proportional to the partial derivative of the particle number flux (2.6) and,

$$\int_{-\infty}^{+\infty} (m_\alpha \boldsymbol{v}) \frac{\partial f_\alpha}{\partial t}\, d\boldsymbol{v} = m_\alpha \frac{\partial}{\partial t}(n_\alpha \langle \boldsymbol{v} \rangle) = m_\alpha \frac{\partial}{\partial t}(n_\alpha \boldsymbol{u}_\alpha)$$

The second addition on the left hand of equation (3.20) is transformed using the following vector identity,

$$\nabla_{\mathbf{r}} \cdot [\boldsymbol{v} \otimes (\boldsymbol{v} f_\alpha)] = [\boldsymbol{v} \cdot \nabla_{\mathbf{r}}] (\boldsymbol{v} f_\alpha) + \cancelto{0}{[(f_\alpha \boldsymbol{v}) \cdot \nabla_{\mathbf{r}}]\, \boldsymbol{v}} = (\boldsymbol{v} \cdot \nabla_{\mathbf{r}})(\boldsymbol{v} f_\alpha)$$

and we have,

$$\begin{aligned} \int_{-\infty}^{+\infty} (m_\alpha \boldsymbol{v})\, \nabla_{\mathbf{r}} \cdot (\boldsymbol{v} f_\alpha)\, d\boldsymbol{v} &= \nabla_{\mathbf{r}} \cdot \left[\int_{-\infty}^{+\infty} m_\alpha (\boldsymbol{v} \otimes \boldsymbol{v}) f_\alpha \, d\boldsymbol{v} \right] \\ &= \nabla_{\mathbf{r}} \cdot (m_\alpha n_\alpha \langle \boldsymbol{v} \otimes \boldsymbol{v} \rangle) \end{aligned}$$

where we can substitute the *momentum flux tensor* $\boldsymbol{S}_\alpha(\boldsymbol{r}, t) = m_\alpha\, n_\alpha \langle \boldsymbol{v} \otimes \boldsymbol{v} \rangle$ of equation (3.15).

The integral of the third term on the left side of equation (3.20) can be also transformed using a vector identity and $(\boldsymbol{F}_{e\alpha} f_\alpha \cdot \nabla_{\mathbf{v}})\, \boldsymbol{v} = f_\alpha\, \boldsymbol{F}_{e\alpha}$ as,

$$\nabla_{\mathbf{v}} \cdot [(\boldsymbol{F}_{e\alpha} f_\alpha) \otimes \boldsymbol{v}] = (\boldsymbol{v} \cdot \nabla_{\mathbf{v}})\, (\boldsymbol{F}_{e\alpha} f_\alpha) + (\boldsymbol{F}_{e\alpha} f_\alpha \cdot \nabla_{\mathbf{v}})\, \boldsymbol{v} = \boldsymbol{v}\, \nabla_{\mathbf{v}} \cdot (\boldsymbol{F}_{e\alpha} f_\alpha) + f_\alpha\, \boldsymbol{F}_{e\alpha}$$

and the third integral in equation (3.20) now reads,

$$\int_{-\infty}^{+\infty} \boldsymbol{v} \nabla_{\mathbf{v}} \cdot (\boldsymbol{F}_{e\alpha} f_\alpha)\, d\boldsymbol{v} = \cancelto{0}{\int_{-\infty}^{+\infty} \nabla_{\mathbf{v}} \cdot [f_\alpha (\boldsymbol{v} \otimes \boldsymbol{F}_{e\alpha})]\, d\boldsymbol{v}} - \int_{-\infty}^{+\infty} f_\alpha \boldsymbol{F}_{e\alpha}\, d\boldsymbol{v}$$

Using the divergence theorem, the first addition on the right hand is null assuming the distribution function f_α decreases monotonically as the velocity $\boldsymbol{v}$ grows. Only the second term remains and for the Lorentz force action on the α-particles we have,

$$\int_{-\infty}^{+\infty} f_\alpha\, \boldsymbol{F}_{e\alpha}\, d\boldsymbol{v} = q_\alpha\, \boldsymbol{E} \left(\int_{-\infty}^{+\infty} f_\alpha\, d\boldsymbol{v} \right) + q_\alpha \left(\left[\int_{-\infty}^{+\infty} \boldsymbol{v} f_\alpha\, d\boldsymbol{v} \right] \times \boldsymbol{B} \right)$$

Using again equations (2.4) and (2.5) we can write,

$$n_\alpha \langle \boldsymbol{F}_{e\alpha} \rangle = q_\alpha\, n_\alpha [\boldsymbol{E} + (\boldsymbol{u}_\alpha \times \boldsymbol{B})] = \rho_{e\alpha}\, \boldsymbol{E} + \boldsymbol{J}_{e\alpha} \times \boldsymbol{B}$$

where $\rho_{e\alpha} = q_\alpha\, n_\alpha$ is the electric charge density and $\boldsymbol{J}_{e\alpha} = q_\alpha\, n_\alpha\, \boldsymbol{u}_\alpha$ electric current density of α-particles that also appear in the Maxwell equations.

With these changes the equation for momentum transport derived from the moment (3.20) now reads,

$$m_\alpha \frac{\partial}{\partial t}(n_\alpha \boldsymbol{u}_\alpha) + \nabla_{\mathbf{r}} \cdot \boldsymbol{S}_\alpha = \rho_{e\alpha}\, \boldsymbol{E} + \boldsymbol{J}_{e\alpha} \times \boldsymbol{B} + \boldsymbol{R}_\alpha \qquad (3.22)$$

We can further transform this equation using divergence of the momentum flux tensor $\boldsymbol{S}_\alpha$ (3.15) as,

$$\nabla_{\mathbf{r}} \cdot \boldsymbol{S}_\alpha = \nabla_{\mathbf{r}} \cdot [n_\alpha\, (\boldsymbol{u}_\alpha \otimes \boldsymbol{u}_\alpha)] + \nabla_{\mathbf{r}} \cdot (p_\alpha\, \boldsymbol{I}) + \nabla_{\mathbf{r}} \cdot \boldsymbol{\Pi}_\alpha$$

and,

$$\nabla_{\mathrm{r}} \cdot \boldsymbol{S}_\alpha = n_\alpha(\boldsymbol{u}_\alpha \cdot \nabla_{\mathrm{r}})\, \boldsymbol{u}_\alpha + \boldsymbol{u}_\alpha\, [\nabla_{\mathrm{r}} \cdot (n_\alpha \boldsymbol{u}_\alpha)] + \nabla_{\mathrm{r}}\, p_\alpha + \nabla_{\mathrm{r}} \cdot \boldsymbol{\Pi}_\alpha$$

Substituting in equation (3.22) we have,

$$\begin{aligned} &m_\alpha n_\alpha \frac{\partial \boldsymbol{u}_\alpha}{\partial t} + m_\alpha \boldsymbol{u}_\alpha \left[\frac{\partial n_\alpha}{\partial t} + \nabla_{\mathrm{r}} \cdot (n_\alpha \boldsymbol{u}_\alpha)\right] \\ &+ m_\alpha n_\alpha (\boldsymbol{u}_\alpha \nabla_{\mathrm{r}} \cdot\,)\, \boldsymbol{u}_\alpha + \nabla_{\mathrm{r}}\, p_\alpha + \nabla_{\mathrm{r}} \cdot \boldsymbol{\Pi}_\alpha = (\rho_{e\alpha}\, \boldsymbol{E} + \boldsymbol{J}_{e\alpha} \times \boldsymbol{B} + \boldsymbol{R}_\alpha) \end{aligned} \tag{3.23}$$

The square brackets contain the continuity equation (3.19) and,

$$m_\alpha \boldsymbol{u}_\alpha \left[\frac{\partial n_\alpha}{\partial t} + \nabla_{\mathrm{r}} \cdot (n_\alpha \boldsymbol{u}_\alpha)\right] = m_\alpha\, \boldsymbol{u}_\alpha\, (S_\alpha - L_\alpha)$$

this contribution accounts for the momentum change associated with the creation/destruction of α-particles. For example, a pair electron/ion is produced in the ionization event of a neutral atom, consequently the ion ($\alpha = i$) and electron ($\alpha = e$) *fluids* experience a change in momentum.

Finally, the momentum transport equation[7] takes a simpler equivalent form and for the ($\alpha = e, i, a$) plasma particles is,

$$\begin{aligned} m_\alpha\, n_\alpha \frac{D\boldsymbol{u}_\alpha}{Dt} &= -\nabla_{\mathrm{r}}\, p_\alpha - \nabla_{\mathrm{r}} \cdot \boldsymbol{\Pi}_\alpha - m_\alpha\, \boldsymbol{u}_\alpha\, (S_\alpha - L_\alpha) + (\rho_{e\alpha}\, \boldsymbol{E} + \boldsymbol{J}_{e\alpha} \times \boldsymbol{B}) \\ &+ \boldsymbol{R}_\alpha \end{aligned} \tag{3.24}$$

where the D/Dt operator is used again. Equation (3.24) is equivalent to the transport equation for macroscopic fluid with electric charge density $\rho_{e\alpha}(\boldsymbol{r}, t) = q_\alpha\, n_\alpha(\boldsymbol{r}, t)$ and current density $\boldsymbol{J}_{e\alpha} = n_\alpha\, q_\alpha\, \boldsymbol{u}_\alpha(\boldsymbol{r}, t)$.

This equation of motion states that the time evolution of fluid momentum for each fluid element of ($\alpha = a, e, i$) particles is due to external electromagnetic forces applied to the fluid, pressure and shear (viscosity) forces. In addition, the collisional interactions $\boldsymbol{R}_\alpha$ as well as the collisional creation and/or recombination of particles through $m_\alpha \boldsymbol{u}_\alpha\, (S_\alpha - L_\alpha)$ also contribute to the momentum.

Specifically, the motions of electrons, ions and neutrals are coupled through the vector $\boldsymbol{R}_\alpha$ which accounts for the momentum exchange in short-range collisions with neutral atoms and also long-range Coulomb encounters between charged particles. Its evaluation requires explicit expressions for $C_\alpha(f_\alpha)$ and the corresponding cross sections for each momentum exchange collision at the microscopic level discussed in volume 1, section 5.4. We postpone this task to section 3.1.3 where a simplified model for $\boldsymbol{R}_\alpha$ essentially valid for short-range collisions will be introduced.

This is also the case of tensor $\boldsymbol{\Pi}_\alpha$ where the calculation of averages $\langle c_i\, c_j \rangle$ in equation (3.14) also involves the velocity distribution function. We face again the problem that the mathematical closure of the macroscopic transport equation (3.24)

[7] For readers who are not familiar with this *convected* time derivative, it is introduced in appendix A.

requires approximations and/or additional physical models, as was indicated in figure 1.1.

Since Boltzmann collision integral is essentially valid for *binary* collisions, the long-range Coulomb interaction between charged particles requires Fokker–Planck or Landau collision operators [2–4].

3.1.3 The friction force

So far we have no expression for the integral of the Boltzmann operator (3.21) which gives $\boldsymbol{R}_\alpha$ in the momentum transport equation. This calculation illustrates the connection between the macroscopic fluid description of plasmas with the properties of elementary processes at the atomic and molecular level discussed in chapters 5 and 6 of volume 1.

The vector $\boldsymbol{R}_\alpha$ accounts for the change in momentum by time and volume units gained or lost by the α-particles in the elastic binary collisions with the other β species,

$$\boldsymbol{R}_\alpha = \sum_{\alpha \neq \beta} \boldsymbol{R}_{\alpha,\beta} \quad \text{where,} \quad \boldsymbol{R}_{\alpha,\beta} = \int_{-\infty}^{+\infty} (m_\alpha \, \boldsymbol{v}) \, C_{\alpha,\beta}(f_\alpha, f_\beta) \, d\boldsymbol{v} \tag{3.25}$$

where $C_{\alpha,\beta}(f_\alpha, f_\beta)$ is given by equation (2.16). Equivalently, equation (3.25) represents the momentum gain/loss rate when α-particles are added or removed from the phase space volume $d^3v \, d^3r$ located at point $(\boldsymbol{r}, t)$ due to collisions with the β species by time unit.

Momentum is conserved in elastic collisions and its change in one particle is equal and opposite to the other colliding particle. Hence, the encounters between α-particles give $\boldsymbol{R}_{\alpha\alpha} = 0$ and are excluded from the summation (3.25) as they do not increase or decrease the average momentum. This is not the case for encounters between different plasma species, for example, a net momentum is transferred from the electron gas to the neutral gas by electron-neutral collisions and then, $\boldsymbol{R}_\alpha \neq 0$. In the general case, the encounters between particles of different species need to be considered in momentum transport equation. Additionally,

$$\boldsymbol{R}_{\alpha\beta} = -\boldsymbol{R}_{\beta\alpha} \quad \text{and,} \quad \sum_\alpha \boldsymbol{R}_\alpha = 0$$

since the amount of momentum gained by one particle group is lost by the others.

In order to derive an expression for $\boldsymbol{R}_{\alpha\beta}$ we consider the elastic collisions between the $dn_\beta = f_\beta(\boldsymbol{v}, \boldsymbol{r}, t) \, d\boldsymbol{v}_\beta$ targets and the $dn_\alpha = f_\alpha(\boldsymbol{v}, \boldsymbol{r}, t) \, d\boldsymbol{v}_\alpha$ incident particles. As discussed in sections[8] 5.1 and 5.4 of volume 1, the number of collision events with only *one* β target with the incoming flow $d\Gamma_\alpha = |\boldsymbol{v}_\alpha - \boldsymbol{v}_\beta| \, dn_\alpha$ (see volume 1, equation (5.3)) of α-particles in the time interval δt is,

[8] The distribution function normalization was different in sections 5.1 and 5.4 of volume 1, where $dn = n_o f(\boldsymbol{u}) \, d\boldsymbol{u}$, so $\int_{-\infty}^{+\infty} f(\boldsymbol{u}) \, d\boldsymbol{u} = 1$, whereas here we use the definition (2.1) usual in kinetic theory and the integral over velocities gives the particle density (2.4).

$$[\,\sigma_{el}(g,\chi)\sin\chi\; d\chi\; d\phi\,] \times \left(\,|v_\alpha - v_\beta|\; dn_\alpha\,\right) \times \delta t$$

Here $\mathbf{g} = v_\alpha - v_\beta$ is the relative velocity and χ the scattering angle (see volume 1, section 5.4 and figure 5.3(b)). The momentum exchanged in one collision along the direction $\boldsymbol{g}$ is given by equation (5.19) in volume 1 as,

$$\Delta \boldsymbol{p} = \mu_{\alpha\beta}\, g\,(1 - \cos\chi) \times \frac{\boldsymbol{g}}{g} = \mu_{\alpha\beta}\,(1 - \cos\chi)\,\boldsymbol{g}$$

where $\mu_{\alpha\beta} = m_\alpha\, m_\beta(m_\alpha + m_\beta)$ is the reduced mass. The momentum exchanged with *one* β target is the number of collision events with α-particles multiplied by $\Delta \boldsymbol{p}$,

$$\left(\,\mu_{\alpha\beta}\, g\,(1 - \cos\chi)\,\boldsymbol{g}\,\right) \times [\,\sigma_{el}(g,\chi)\sin\chi\; d\chi\; d\phi\,] \times \left(\,|v_\alpha - v_\beta|\; dn_\alpha\,\right) \times \delta t$$

When dn_β targets are considered, the momentum exchange rate by time and volume units $\delta \boldsymbol{P}/\delta t$ is given by,

$$\frac{\delta \boldsymbol{P}}{\delta t} = \Delta \boldsymbol{p} \times \dot{Q}_{el} = \left(\mu_{\alpha\beta}\,(1 - \cos\chi)\,\boldsymbol{g}\,\right)$$
$$\times[\,\sigma_{el}(g,\chi)\sin\chi\; d\chi\; d\phi\,] \times \left(|v_\alpha - v_\beta|\; dn_\alpha\,\right) \times dn_\beta$$

Here $\dot{Q}_{el}$ is the *collision rate*, the number of elastic collisions by time unit introduced in chapter 5 of volume 1. Since the scattering angles $(\chi,\ \phi)$ are independent of v_α and v_β, the integration gives,

$$\frac{\delta \boldsymbol{P}}{\delta t} = \delta \boldsymbol{p} \times \dot{Q}_{el} = \left[2\pi \int_0^{\pi} \sigma_{el}(g,\chi)\,(1 - \cos\chi)\sin\chi\; d\chi\right]$$
$$\times\, \mu_{\alpha\beta}\, \boldsymbol{g}\; g\; (f_\alpha\; dv_\alpha)\,(f_\beta\; dv_\beta)$$

The expression between the brackets $\sigma_{tr}(g)$ is the *momentum transfer cross section* for elastic collisions (volume 1, equation (5.20)). Finally, the change in momentum by time and volume units $\boldsymbol{R}_{\alpha\beta}$ is obtained by integration over the velocities v_α and v_β

$$\boldsymbol{R}_{\alpha\beta} = \mu_{\alpha\beta} \int_{-\infty}^{+\infty} \int_{-\infty}^{+\infty} \sigma_{tr}(g)\, \boldsymbol{g}\; g\, f_\alpha\, f_\beta\; dv_\alpha\; dv_\beta$$

This last equation explicitly shows the connection between the physical characteristics of collisions, considered in the momentum transfer cross section $\sigma_{tr}(g)$, and the momentum transport equation (3.24). The electron cross sections of figure 6.3 in volume 1 with neutral gas atoms give the experimental values for $\sigma_{tr}(g)$ that are connected to the macroscopic magnitude $\boldsymbol{R}_{ea}$. This illustrates the influence of the specific properties of collisions at the atomic and molecular level in the macroscopic transport of momentum and energy.

However, the calculation of $\boldsymbol{R}_{\alpha\beta}$ needs, in addition to the cross section $\sigma_{tr}(g)$, the explicit expressions for the velocity distribution functions for the α and β species that are not frequently available. The usual simplifying assumption is to consider equilibrium Maxwellian distributions as in section 3.4 of volume 1 for the calculation of the rate constants. Furthermore, the cross section $\sigma_{tr}(g) \simeq \sigma_{tr}$ is considered constant and independent of the relative velocity. The calculation is cumbersome and it will not be reproduced here, the interested reader can find the details in section 6.5 of reference [3]. The result is,

$$\boldsymbol{R}_{\alpha\beta} = -\mu_{\alpha\beta}\, n_\alpha\, \nu_{\alpha\beta}\, (\boldsymbol{u}_\alpha - \boldsymbol{u}_\beta) \tag{3.26}$$

where $\nu_{\alpha\beta}$ is the collision frequency for momentum transfer given by,

$$\nu_{\alpha\beta} = n_\beta\, \sigma_{tr}\, \frac{4}{3}\left(\frac{8\, k_B\, T}{\pi\, \mu_{\alpha\beta}}\right)$$

In equation (3.26) the collisional interaction between plasma species can be seen as a *friction* or *drag force* $\boldsymbol{R}_{\alpha\beta}$ caused by their relative motion of plasma species.

This expression (3.26) for the friction force is useful for the elastic collisions of electrons ($\mu_{ea} \simeq m_e$) with ions and neutrals. It is also valid for elastic encounters between heavy particles, however, a more involved collision operator is required to account for charge exchange collisions discussed in section 6.3 of volume 1.

3.1.4 The energy transport equation

The next moment of the Boltzmann equation, related to the energy transport is obtained by multiplying equation (2.10) by $m_\alpha v^2/2$ and integrating over $\boldsymbol{v}$ as,

$$\begin{aligned}\frac{\partial}{\partial t}\left(n_\alpha\left\langle\frac{m_\alpha v^2}{2}\right\rangle\right) + \nabla_{\mathrm{r}}\cdot\left(n_\alpha\left\langle\frac{m_\alpha v^2}{2}\,\boldsymbol{v}\right\rangle\right)& \\ + \frac{1}{2}\int_{-\infty}^{+\infty} v^2\, \nabla_{\mathrm{v}}\cdot(\boldsymbol{F}_{e\alpha} f_\alpha)\, d\boldsymbol{v} &= \int_{-\infty}^{+\infty}\frac{m_\alpha v^2}{2}\, C_s(f_\alpha)\, d\boldsymbol{v}\end{aligned} \tag{3.27}$$

where $\boldsymbol{F}_{e\alpha} = q_\alpha\,[\boldsymbol{E} + (\boldsymbol{v}\times\boldsymbol{B})]$ is again the Lorentz force acting on the α particle. The first term on the left side of equation (3.27) is the time derivative of the energy by volume unit introduced in equation (3.10) as,

$$\frac{\partial}{\partial t}\left(n_\alpha\left\langle\frac{m_\alpha v^2}{2}\right\rangle\right) = \frac{\partial}{\partial t}\left(m_\alpha\, n_\alpha\,\frac{u_\alpha^2}{2} + e_{i\alpha}\right) = \frac{\partial E_\alpha}{\partial t} \tag{3.28}$$

The right term (3.27) involves the collision operator $C_s(f_\alpha)$ as,

$$Q_\alpha = \int_{-\infty}^{+\infty}\frac{m_\alpha v^2}{2}\, C_s(f_\alpha)\, d\boldsymbol{v} \tag{3.29}$$

that will be discussed later and represents the rate of energy density change due to collisions.

The energy flux vector $\boldsymbol{K}_\alpha(\boldsymbol{r}, t)$ of equation 3.18 can be introduced in the second average of equation (3.27) and,

$$\nabla_{\mathrm{r}}\cdot\left(n_\alpha\left\langle\frac{m_\alpha v^2}{2}\,\boldsymbol{v}\right\rangle\right) = \nabla_{\mathrm{r}}\cdot(E_\alpha\,\boldsymbol{u}_\alpha) + \nabla_{\mathrm{r}}\cdot\boldsymbol{q}_\alpha + \nabla_{\mathrm{r}}\cdot(p_\alpha\,\boldsymbol{u}_\alpha) + \nabla_{\mathrm{r}}\cdot(\boldsymbol{\Pi}_\alpha\colon\boldsymbol{u}_\alpha) \tag{3.30}$$

The third term on the left side of equation (3.27) can be transformed as,

$$\nabla_v \cdot (\boldsymbol{F}_{ea}\, v^2 f_\alpha) = v^2\, \nabla_v \cdot (\boldsymbol{F}_{ea} f_\alpha) + (f_\alpha\, \boldsymbol{F}_{ea}) \cdot (\nabla_v\, v^2)$$

Using the divergence theorem we obtain again for a well behaved distribution as $|v| \to \infty$,

$$\int_{-\infty}^{+\infty} v^2\, \nabla_{\mathbf{v}} \cdot (\boldsymbol{F}_{ea} f\alpha)\, d\boldsymbol{v} = \int_{-\infty}^{+\infty} \cancelto{0}{\nabla_{\mathbf{v}} \cdot (v^2 \boldsymbol{F}_{ea} f\alpha)\, d\boldsymbol{v}} - \int_{-\infty}^{+\infty} (\boldsymbol{F}_{ea} f\alpha) \cdot \nabla_{\mathbf{v}}(v^2)\, d\boldsymbol{v}$$

In the above equation the first integral is null because f_α decreases as v increases and using $\nabla_v\, (v^2) = 2\, \boldsymbol{v}$ we finally have,

$$\frac{1}{2} \int_{-\infty}^{+\infty} v^2\, \nabla_v \cdot (\boldsymbol{F}_{ea} f_\alpha)\, dv = -\frac{1}{2} \int_{-\infty}^{+\infty} (\boldsymbol{F}_{ea} f_\alpha) \cdot (2\, \boldsymbol{v})\, dv = -\, n_\alpha\, \langle \boldsymbol{F}_{ea} \cdot \boldsymbol{v} \rangle$$

and the last term in equation (3.27) is the work of the electric field,

$$n_\alpha\, \langle \boldsymbol{F}_{ea} \cdot \boldsymbol{v} \rangle = q_\alpha\, \boldsymbol{E}(\boldsymbol{r}, t) \cdot \left(n_\alpha \int_{-\infty}^{+\infty} f_\alpha\, \boldsymbol{v}\, dv \right) = \boldsymbol{E} \cdot (q_\alpha\, n_\alpha\, \boldsymbol{u}_\alpha) = \boldsymbol{E} \cdot \boldsymbol{J}_{ea} \tag{3.31}$$

Finally, substituting in equation (3.27) equations (3.28)–(3.31) we obtain the energy transport equation,

$$\begin{aligned} \left[\frac{\partial E_\alpha}{\partial t} + \nabla_r \cdot (E_\alpha\, \boldsymbol{u}_\alpha) \right] = &- \nabla_r \cdot \boldsymbol{q}_\alpha - \nabla_r \cdot (p_\alpha\, \boldsymbol{u}_\alpha) \\ &- \nabla_r \cdot (\boldsymbol{\Pi}_\alpha \colon \boldsymbol{u}_\alpha) + \boldsymbol{J}_{ea} \cdot \boldsymbol{E} + Q_\alpha \end{aligned} \tag{3.32}$$

The left side expresses a continuity equation for the energy E_α in a small fluid element and the right terms are the gain and loss terms. The first right addition is the variation in E_α due to the heat flux $\boldsymbol{q}_\alpha$ and next are the work done by the kinetic pressure $\boldsymbol{P}_\alpha = p_\alpha\, \boldsymbol{I} + \boldsymbol{\Pi}_\alpha$ composed of the scalar pressure p_α and viscosity terms The term $\boldsymbol{J}_{ea} \cdot \boldsymbol{E}$ is the work of current transport in the electric field, whereas Q_α accounts for the collisional energy exchanges.

It is convenient to rewrite equation (3.32) in terms of the pressure p_α or temperature T_α, we expand all terms,

$$\begin{aligned} &\frac{m_\alpha\, u_\alpha^2}{2} \frac{\partial n_\alpha}{\partial t} + m_\alpha\, n_\alpha\, \boldsymbol{u}_\alpha \cdot \frac{\partial \boldsymbol{u}_\alpha}{\partial t} + \frac{\partial e_{i\alpha}}{\partial t} + \nabla_r \cdot (e_{i\alpha}\, \boldsymbol{u}_\alpha) = -\nabla_r \cdot (p_\alpha\, \boldsymbol{u}_\alpha) \\ &- \frac{m_\alpha\, u_\alpha^2}{2} \nabla_r \cdot (n_\alpha\, \boldsymbol{u}_\alpha) - \frac{m_\alpha}{2} (n_\alpha\, \boldsymbol{u}_\alpha) \cdot \nabla_r\, (u_\alpha^2) - \nabla_r \cdot \boldsymbol{q}_\alpha - \nabla_r \cdot (\boldsymbol{\Pi}_\alpha \colon \boldsymbol{u}_\alpha) + \boldsymbol{J}_\alpha \cdot \boldsymbol{E} + Q_\alpha \end{aligned}$$

In the second addition in the second line $\boldsymbol{u}_\alpha \cdot \nabla_r\, (u_\alpha^2) = \boldsymbol{u}_\alpha \cdot [2\, (\boldsymbol{u}_\alpha \cdot \nabla_r)\, \boldsymbol{u}_\alpha]$ and we obtain,

$$\frac{m_\alpha u_\alpha^2}{2}\left[\frac{\partial n_\alpha}{\partial t} + \nabla_r \cdot (n_\alpha \boldsymbol{u}_\alpha)\right] + \boldsymbol{u}_\alpha \cdot \left[m_\alpha n_\alpha\left(\frac{\partial \boldsymbol{u}_\alpha}{\partial t} + (\boldsymbol{u}_\alpha \cdot \nabla_r)\, \boldsymbol{u}_\alpha\right)\right]$$
$$+ \frac{\partial e_{i\alpha}}{\partial t} + \nabla_r \cdot (e_{i\alpha} \boldsymbol{u}_\alpha) + \nabla_r \cdot (p_\alpha \boldsymbol{u}_\alpha) + \nabla_r \cdot \boldsymbol{q}_\alpha + \nabla_r \cdot (\boldsymbol{\Pi}_\alpha\colon \boldsymbol{u}_\alpha) = \boldsymbol{J}_\alpha \cdot \boldsymbol{E} + Q_\alpha$$

Between the square brackets are the continuity equation (3.19) and the moment transport equation (3.24) so,

$$\frac{m_\alpha u_\alpha^2}{2}(S_\alpha - L_\alpha)$$
$$+ \boldsymbol{u}_\alpha \cdot \left[-\nabla_r p_\alpha - \nabla_r \cdot \boldsymbol{\Pi}_\alpha - m_\alpha \boldsymbol{u}_\alpha (S_\alpha - L_\alpha) + (\rho_{e\alpha} \boldsymbol{E} + \boldsymbol{J}_{e\alpha} \times \boldsymbol{B}) + \boldsymbol{R}_\alpha\right]$$
$$+ \frac{\partial e_{i\alpha}}{\partial t} + \nabla_r \cdot (e_{i\alpha} \boldsymbol{u}_\alpha) + \nabla_r \cdot (p_\alpha \boldsymbol{u}_\alpha) + \nabla_r \cdot \boldsymbol{q}_\alpha + \nabla_r \cdot (\boldsymbol{\Pi}_\alpha\colon \boldsymbol{u}_\alpha) = \boldsymbol{J}_\alpha \cdot \boldsymbol{E} + Q_\alpha$$

Using the identity $\nabla_r \cdot (\boldsymbol{\Pi}_\alpha\colon \boldsymbol{u}_\alpha) = \boldsymbol{u}_\alpha \cdot (\nabla_r \cdot \boldsymbol{\Pi}_\alpha) + (\boldsymbol{\Pi}_\alpha \cdot \nabla_r)\, \boldsymbol{u}_\alpha$ and after a little algebra we obtain,

$$\frac{De_{i\alpha}}{Dt} + (e_{i\alpha} + p_\alpha)(\nabla_r \cdot \boldsymbol{u}_\alpha) + (\boldsymbol{\Pi}_\alpha \cdot \nabla_r)\, \boldsymbol{u}_\alpha + \nabla_r \cdot \boldsymbol{q}_\alpha$$
$$= Q_\alpha - \boldsymbol{u}_\alpha \cdot \boldsymbol{R}_\alpha + \frac{m_\alpha u_\alpha^2}{2}(S_\alpha - L_\alpha)$$

Using the LTE approximation (3.11) this equation can be expressed in terms of the temperature as $e_{i\alpha} = 3n_\alpha k_B T_\alpha/2$ or of the scalar pressure $e_{i\alpha} = 3p_\alpha/2$. In this last case we finally obtain,

$$\frac{3}{2}\frac{Dp_\alpha}{Dt} + \frac{5}{2} p_\alpha (\nabla_r \cdot \boldsymbol{u}_\alpha) + (\boldsymbol{\Pi}_\alpha \cdot \nabla_r)\, \boldsymbol{u}_\alpha = -\nabla_r \cdot \boldsymbol{q}_\alpha$$
$$- \boldsymbol{u}_\alpha \cdot \boldsymbol{R}_\alpha + \frac{m_\alpha u_\alpha^2}{2}(S_\alpha - L_\alpha) + Q_\alpha \tag{3.33}$$

Here $-\boldsymbol{u}_\alpha \cdot \boldsymbol{R}_\alpha$ is the work of the friction force, the next term accounts for the kinetic energy of created/destroyed α-particles. The term Q_α in equation (3.34c) is the second order moment (3.29),

$$Q_\alpha = \int_{-\infty}^{+\infty} \frac{m_\alpha v^2}{2}\, C_s(f_\alpha)\, dv$$

and again, its rigorous calculation is cumbersome so we will make use of approximate expressions. The term Q_α accounts for the energy transformed in inelastic collisions. For example, the ionizing electron loses an amount of energy E_I equal to the ionization potential of the neutral gas and when the new ion is created at rest with respect to the ion fluid we can write,

$$Q_I = E_i\, \nu_I$$

where ν_i is the ionization frequency and is given by, $\nu_I\, n_e = k_I\, n_a\, n_e$ introduced in chapter 6 of volume 1.

Equation (3.33) shows again the connection between the elementary processes at the molecular and atomic level with the macroscopic description of a plasma. Finally, the heat flux vector $\boldsymbol{q}_\alpha$, the collisional energy exchange Q_α between plasma species and the components of the tensor $\boldsymbol{P}_\alpha = p_\alpha\, \boldsymbol{I} + \boldsymbol{\Pi}_\alpha$ still remain undetermined in the above energy transport equation. Again, their calculation requires an expression of the velocity distribution function $f_\alpha(\boldsymbol{v}, \boldsymbol{r}, t)$ and/or additional approximations. These undetermined quantities are also related to the *closure problem* introduced in section 3.2.1.

3.2 The hydrodynamic plasma transport equations

The above fluid transport equations (3.19), (3.24) and (3.32) for a plasma ($\alpha = e, i, a$) are,

$$\frac{\partial n_\alpha}{\partial t} + \nabla_{\mathrm{r}} \cdot (n_\alpha\, \boldsymbol{u}_\alpha) = S_\alpha - L_\alpha \tag{3.34a}$$

$$m_\alpha\, n_\alpha \frac{D\boldsymbol{u}_\alpha}{Dt} = -\nabla_{\mathrm{r}}\, p_\alpha - \nabla_{\mathrm{r}} \cdot \boldsymbol{\Pi}_\alpha - m_\alpha\, \boldsymbol{u}_\alpha\, (S_\alpha - L_\alpha) + \boldsymbol{F}_{e\alpha} + \boldsymbol{R}_\alpha \tag{3.34b}$$

$$\begin{aligned} &\frac{3}{2}\frac{Dp_\alpha}{Dt} + \frac{5}{2}\, p_\alpha\, (\nabla_{\mathrm{r}} \cdot \boldsymbol{u}_\alpha) + (\boldsymbol{\Pi}_\alpha \cdot \nabla_{\mathrm{r}})\, \boldsymbol{u}_\alpha \\ &\quad + \nabla_{\mathrm{r}} \cdot \boldsymbol{q}_\alpha = Q_\alpha - \boldsymbol{u}_\alpha \cdot \boldsymbol{R}_\alpha + \frac{m_\alpha\, u_\alpha^2}{2}\, (S_\alpha - L_\alpha) \end{aligned} \tag{3.34c}$$

where $\boldsymbol{F}_{e\alpha} = (\rho_{e\alpha}\, \boldsymbol{E} + \boldsymbol{J}_{e\alpha} \times \boldsymbol{B})$ with $\rho_{e\alpha} = q_\alpha\, n_\alpha$ and $\boldsymbol{j}_{e\alpha} = n_\alpha\, q_\alpha\, \boldsymbol{u}_\alpha$ are the electromagnetic forces. The vector $\boldsymbol{R}_\alpha$ is the sum of the *friction forces* $\boldsymbol{R}_\alpha = \sum_{\alpha\neq\beta} \boldsymbol{R}_{\alpha\beta}$ introduced in section 3.1.3. In the equilibrium of particle production and losses $S_\alpha = L_\alpha$ and some terms in equations (3.34) are null.

Equations (3.34) are also called *multifluid plasma equations* and have been derived from the Boltzmann equation (2.10) under the assumption of the LTE *local equilibrium* introduced in section 4.2 of volume 1. These are collision-dominated plasmas where fast energy relaxation processes relax fluctuations on time and length $l_c \ll L_s$ and time $\tau \ll L_s/|\boldsymbol{u}_\alpha|$ scales much shorter than the macroscopic plasma motion L_s as discussed in the introductory section.

The plasma medium can be also considered as a *conducting fluid* and equations (3.34) can be combined to describe its macroscopic behavior *as a whole* without regarding the individual motions of its constituent species. The parameters of interest are such as the total mass density as $\rho_m = \sum_\alpha \rho_{m\,\alpha}$ or the mean fluid velocity $\rho_m\, \boldsymbol{u} = \sum_\alpha \rho_{m\,\alpha}\, \boldsymbol{u}_\alpha$ where the contribution of each plasma species is weighted. In order to derive the hydrodynamic equations for an electrically conductive single fluid, the previous formulation of transport equations needs to be modified accordingly.

The resulting transport equations for a conducting fluid are called the *magneto-hydrodynamic equations* (MHD) and are not frequently used in their complex general form. Several approximations are usually considered to eliminate some terms and to reduce the MHD equations to a simplified equation set. However, these tasks are beyond the scope of the present book.

3.2.1 The closure of the transport equations

Unfortunately, the equations (3.34) have undetermined coefficients related with the moments of the distribution function $f_\alpha(v, \boldsymbol{r}, t)$ that need to be supplied by additional physical models and/or approximations.

The first moment of the Boltzmann equation (2.10) gives the continuity equation (3.34a) that relates the number density $n_\alpha(\boldsymbol{r}, t)$ with the fluid velocity $\boldsymbol{u}_\alpha(\boldsymbol{r}, t)$. Since these are two independent macroscopic variables we need an additional moment to derive the momentum conservation equation (3.34b). However, this relates the fluid velocity with the kinetic stress tensor $\boldsymbol{P}_\alpha = p_\alpha \boldsymbol{I} + \boldsymbol{\Pi}_\alpha$ and equations (3.34a) and (3.34b) become two equations with three independent macroscopic variables. The following moment gives the energy transport equation (3.34c) which involves the heat flux vector $\boldsymbol{q}(\boldsymbol{r}, t)$ in addition to kinetic pressure $\boldsymbol{S}_\alpha$, the number density $n_\alpha(\boldsymbol{r}, t)$ and the mean velocity $\boldsymbol{u}_\alpha(\boldsymbol{r}, t)$.

Consequently, any *finite* number of moments of the Boltzmann equation (2.10) lead a set of fluid equations with more unknowns than equations. This problem of the *mathematical closure* is inherent to macroscopic transport equations and we restrict ourselves to the first three moments since these are related with the magnitudes of physical interest.

Conventional hydrodynamics circumvents this problem introducing phenomenological expressions for the undetermined variables in equation (3.34), For example, the usual expression for the stress tensor $\boldsymbol{\Pi}_\alpha$ or the Fourier law $\boldsymbol{q} = -\kappa\, \nabla_{\mathbf{r}} T$ for the heat flux where T is the local temperature and κ is the thermal conductivity[9] of gases or liquids. These relations constitute the *additional information* mentioned in the scheme of figure 1.1.

However, this procedure is not fully justified in the case of plasmas where similar expressions are not available. The explicit form of the velocity distribution function is required—or equivalently—a specific *plasma model* is required to find expressions for the undetermined variables in the fluid equations. For example, the evaluation of the heat flux vector,

$$q_\alpha = \int_{-\infty}^{+\infty} \left(\frac{m_\alpha c^2}{2}\right) \boldsymbol{c}\, f_\alpha(v, \boldsymbol{r}, t)\, d\boldsymbol{c}$$

requires a specific velocity distribution $f_\alpha(v, \boldsymbol{r}, t)$ (or equivalently, a physical model) for electrons, ions and neutral atoms. The same argument applies to the other indeterminate quantities in equations (3.34).

There are two available schemes to circumvent this difficulty. The simplest choice is to *truncate* the moment expansion, assuming the higher order moments arbitrarily as zero or alternatively, in terms of lower order moments on the basis of physical

[9] In the general case the thermal conductivity is also a tensor.

assumptions. The implicit risk is not to reflect relevant physical phenomena in these simple approaches, such as the *cold* and *warm* plasma models discussed below.

The alternative is *asymptotic schemes* where the mathematical expansion of a small parameter controls the approximations, as for example, the ratio $\epsilon = l_c/L_c \ll 1$ in Chapman–Enskog theory [2, 4] for low pressure neutral gases. This permits expanding the distribution function as,

$$f_\alpha(\boldsymbol{v}, \boldsymbol{r}, t) = f_{o,\alpha}(\boldsymbol{v}, \boldsymbol{r}, t) + f_{1,\alpha}(\boldsymbol{v}, \boldsymbol{r}, t)\,\epsilon + f_{2,\alpha}(\boldsymbol{v}, \boldsymbol{r}, t)\,\epsilon^2 \ldots$$

In this expansion the functions $f_{i,\alpha}(\boldsymbol{v}, \boldsymbol{r}, t)$ with $i = 1, 2, \ldots$ are assumed of the same order of magnitude and reference LTE equilibrium state $f_{o,\alpha}(\boldsymbol{v}, \boldsymbol{r}, t)$ is described by a local Maxwellian distribution (equation (4.3) of volume 1). This asymptotic closure scheme based on the $\epsilon = l_c/L_c \ll 1$ parameter is valid for collisional plasmas where short-range encounters dominate.

The second approach for magnetized plasmas is based on the ratio $\epsilon = R_{l\alpha}/L_c \ll 1$ between the Larmor radius $R_{l\alpha}$ (see section 4.5.4 of volume 1) and the macroscopic scale of plasma motion. The resulting macroscopic equations valid for fully ionized and magnetized plasma flows are called *Braginskii equations* [2, 5]. No asymptotic closure scheme for collisionless unmagnetized plasmas has been formulated so far.

These asymptotic expansion closure approaches are cumbersome and less employed than the truncation schemes discussed below. The interested reader is referred to the specialized literature [2, 4, 5] since this specific subject is outside the scope of the present work.

3.2.2 The cold and warm plasma models

The simplest closed set of hydrodynamic plasma equations is called the *cold plasma* model where the kinetic pressure $\boldsymbol{P}_\alpha = p_\alpha \boldsymbol{I} + \boldsymbol{\Pi}_\alpha$ is neglected. Equations (3.34) reduce in this case to,

$$\frac{\partial n_\alpha}{\partial t} + \nabla_{\mathbf{r}} \cdot (n_\alpha\, \boldsymbol{u}_\alpha) = S_\alpha - L_\alpha \tag{3.35a}$$

$$m_\alpha\, n_\alpha \frac{D\boldsymbol{u}_\alpha}{Dt} = (\rho_{e\alpha}\, \boldsymbol{E} + \boldsymbol{J}_{e\alpha} \times \boldsymbol{B}) + \boldsymbol{R}_\alpha - m_\alpha\, \boldsymbol{u}_\alpha\, (S_\alpha - L_\alpha) \tag{3.35b}$$

In the *cold plasma* approximation the thermal motions of particles are disregarded or equivalently zero kinetic temperatures T_α are considered for the plasma species. Hence, this model considers their velocity distribution functions,

$$f_\alpha(\boldsymbol{v}, \boldsymbol{r}, t) = \delta[\boldsymbol{v} - \boldsymbol{u}_\alpha(\boldsymbol{r}, t)]$$

are Dirac deltas centered at the macroscopic flow velocity.

The conservation of energy is the simplifying approximation introduced in the *warm plasma* model. The terms involving the heat flux vector $\boldsymbol{q}$ or equivalently, the plasma thermal conductivity are considered as null in equations (3.34). Hence the plasma is non-viscous and also the tensor $\boldsymbol{\Pi}_\alpha$ corresponding to the non-diagonal terms of the kinetic pressure are neglected and we have,

$$\frac{\partial n_\alpha}{\partial t} + \nabla_{\mathbf{r}} \cdot (n_\alpha \boldsymbol{u}_\alpha) = S_\alpha - L_\alpha \tag{3.36a}$$

$$m_\alpha n_\alpha \frac{D\boldsymbol{u}_\alpha}{Dt} = -\nabla_{\mathbf{r}} p_\alpha + (\rho_{e\alpha} \boldsymbol{E} + \boldsymbol{J}_{e\alpha} \times \boldsymbol{B}) + \boldsymbol{R}_\alpha - m_\alpha \boldsymbol{u}_\alpha (S_\alpha - L_\alpha) \tag{3.36b}$$

$$\frac{3}{2}\frac{Dp_\alpha}{Dt} + \frac{5}{2} p_\alpha (\nabla_{\mathbf{r}} \cdot \boldsymbol{u}_\alpha) = Q_\alpha - \boldsymbol{u}_\alpha \cdot \boldsymbol{R}_\alpha + \frac{m_\alpha u_\alpha^2}{2} (S_\alpha - L_\alpha) \tag{3.36c}$$

In this approximation, the scalar pressure p_α is preserved to be consistent with the LTE equilibrium conditions as it is proportional to the kinetic temperature T_α of the plasma species (see equations (3.12) and (3.13)).

The equations for a plasma in equilibrium derived in chapter 4 of volume 1 on the basis of the Maxwellian velocity distribution function are recovered from equation (3.36b) as shown in box 3.2. Furthermore, when collisions are neglected, $\boldsymbol{R}_\alpha$, Q_α are null as well as L_α and S_α. In this case the warm plasma model is also denominated *adiabatic approximation* since equation (3.36c) reduces to the $p_\alpha\, \rho_{m\alpha}^{-\gamma}=$ constant, as shown in box 3.3.

Finally, *hybrid* approaches are possible where ions and/or neutrals are considered *cold* since their typical kinetic temperatures are much lower than those of the electron fluid as in section 3.4. These can be considered *warm* compared with the ion fluid background since the thermal motion of the lighter electrons is much faster.

Box 3.2. Plasma equilibrium

In the steady equilibrium of an electron–ion ($\alpha = i$, e and $q_\alpha = \pm e$) plasma under an external electric field $\boldsymbol{E} = -\nabla_{\mathbf{r}}\phi$ with $\boldsymbol{B} = 0$ and equation (3.36b) becomes,

$$m_\alpha n_\alpha \cancelto{0}{\frac{D\boldsymbol{u}_\alpha}{Dt}} = -\nabla_{\mathbf{r}} p_\alpha + \rho_{e\alpha} \boldsymbol{E} = 0 \quad \text{and,} \quad -\nabla_{\mathbf{r}}(n_\alpha k_B T) \pm n_\alpha e(-\nabla_{\mathbf{r}}\phi) = 0$$

The fluid velocity is $\boldsymbol{u}_\alpha = 0$ since the plasma remains at rest and $S_\alpha = L_\alpha$ as the ionization and recombination rates of charged particles are equal. In these conditions,

$$\nabla_{\mathbf{r}}\left[\ln(n_\alpha) \mp \frac{e\,\phi}{k_B\,T}\right] = 0 \quad \text{hence,} \quad \ln(n_\alpha) \mp \frac{e\,\phi}{k_B\,T} = \text{constant}$$

The expression between the brackets is uniform and we recover equations (4.5) and (4.5) of volume 1 for a plasma in equilibrium,

$$n_\alpha = n_o \exp\left(\mp \frac{e\,\phi}{k_B T}\right)$$

These expressions were derived in chapter 4 of volume 1 using the Maxwellian distribution function (4.4) for the velocity distribution of ions and electrons in the Maxwellian plasma equilibrium state.

Box 3.3. The adiabatic approximation

When collisions can be neglected equations (3.36a) and (3.36c) reduce to,

$$\frac{\partial n_\alpha}{\partial t} + \nabla_{\mathbf{r}} \cdot (n_\alpha \, \boldsymbol{u}_\alpha) = 0$$

$$\frac{3}{2}\frac{Dp_\alpha}{Dt} + \frac{5}{2} p_\alpha \, (\nabla_{\mathbf{r}} \cdot \boldsymbol{u}_\alpha) = 0$$

We can use the first equation to substitute $\nabla_{\mathbf{r}} \cdot \boldsymbol{u}_\alpha$ in the second since,

$$\frac{\partial n_\alpha}{\partial t} + \boldsymbol{u}_\alpha \cdot \nabla_{\mathbf{r}} n_\alpha + n_\alpha \nabla_{\mathbf{r}} \cdot \boldsymbol{u}_\alpha = \frac{Dn_\alpha}{Dt} + n_\alpha \nabla_{\mathbf{r}} \cdot \boldsymbol{u}_\alpha = 0$$

We multiply this last equation by the particle mass m_α so $m_\alpha \, n_\alpha = \rho_{m\alpha}$ is the density, then,

$$\frac{3}{2}\frac{1}{p_\alpha}\frac{Dp_\alpha}{Dt} - \frac{5}{2}\frac{1}{\rho_{m\alpha}}\frac{D\rho_{m\alpha}}{Dt} = 0, \qquad \frac{D}{Dt}\left[\frac{3}{2}\ln(p_\alpha) - \frac{5}{2}\ln(\rho_{m\alpha})\right] = 0$$

Finally we obtain the relation $p_\alpha \, \rho_{m\alpha}^{-5/3}$ = constant usual in thermodynamics for the adiabatic expansion of gas with $\gamma = 5/3$, since we made use of the ideal monoatomic gas approximation (3.11).

This applies to PLE plasmas discussed in sections 4.2 and 4.3 of volume 1 so far as the macroscopic local equilibrium is preserved on some length and time scales.

3.2.3 The diffusion approximation

In the steady equilibrium of unmagnetized ($\boldsymbol{B} = 0$) weakly ionized plasmas the production and losses of charged particles are balanced and short-range collisions with neutral atoms background are dominant. We consider a *cold* (T_i, $T_e \gg T_a$) and uniform distribution of neutrals with a macroscopic velocity ($\boldsymbol{u}_a \simeq 0$) much slower than those of charged particles.

In these conditions the friction force (3.26) for ions and electrons can be approximated by $\boldsymbol{R}_{\alpha a} = -\mu_{\alpha a} \, n_\alpha \, \nu_{\alpha a} \, \boldsymbol{u}_\alpha$ since long-range Coulomb collisions between charged particles are negligible. The momentum transport equation (3.36b) for ions and electrons becomes,

$$m_\alpha \, n_\alpha \frac{D\boldsymbol{u}_\alpha}{Dt} = -\nabla_{\mathbf{r}} \, (n_\alpha \, k_B \, T) \pm e \, n_\alpha \, \boldsymbol{E} - \mu_{\alpha a} \, n_\alpha \, \nu_{\alpha a} \, \boldsymbol{u}_\alpha$$

where $\mu_{ia} = m_i/2$ and $\mu_{ea} = m_e$ and $\nu_{\alpha a}$ are the corresponding momentum transfer collision frequencies. When the macroscopic acceleration $D\boldsymbol{u}_\alpha/Dt \simeq 0$ is null the force imparted on charged particles by the electric field is balanced by collisions and pressure forces. Then,

$$\boldsymbol{u}_\alpha = \pm \frac{e}{\mu_{\alpha a}\, \nu_{\alpha a}} \boldsymbol{E} - \frac{k_B\, T}{\mu_{\alpha a}\, \nu_{\alpha a}} \frac{\nabla_\mathbf{r}\, n_\alpha}{n_\alpha} \tag{3.37}$$

the velocity of ions and electrons has two components; one along the gradient $\nabla_\mathbf{r}\, n_\alpha$ and a *drift speed* $\boldsymbol{u}_{d\alpha} = \mu_\alpha\, \boldsymbol{E}$ parallel to the electric field $\boldsymbol{E}$. The coefficient $\mu_\alpha = (e/\mu_{\alpha a}\, \nu_{\alpha a})$ where $\mu_{\alpha,a} = m_a\, m_\alpha/(m_a + m_\alpha)$ is called *mobility* and the second $D_\alpha = (k_B\, T/\mu_{\alpha a}\, \nu_{\alpha a})$ is a *diffusion coefficient*. We can write for the particle flux,

$$\boldsymbol{\Gamma}_\alpha = n_\alpha\, \boldsymbol{u}_\alpha = \pm\, \mu_\alpha\, n_\alpha\, \boldsymbol{E} - D_\alpha\, \nabla_\mathbf{r}\, n_\alpha$$

which reduces to the Fick's law of diffusion when $\boldsymbol{E} = 0$ or also for neutral particles, which are unaffected by the electric field.

This equation expresses that in isothermal plasmas, short-range collision dominating the motion of ions and electrons is driven by the electric field with velocity $v_{d\alpha}$ and the charges also perform a random walk displacement from regions with high concentrations to less dense zones. Since the electric current transported is $\boldsymbol{J}_e = \sigma_c\, \boldsymbol{E}$ the drift speed of charged species ($\alpha = e, i$) contributes to the plasma electric conductivity σ_c.

However, both charged species cannot move separately so $\boldsymbol{\Gamma}_i = \boldsymbol{\Gamma}_e$ and quasineutrality $n_i \simeq n_e = n$ must be preserved, otherwise intense electric fields develop in the plasma. Thus, we can write,

$$\boldsymbol{\Gamma}_i = \mu_i\, n\, \boldsymbol{E} - D_i\, \nabla_\mathbf{r}\, n = \mu_e\, n\, \boldsymbol{E} - D_e\, \nabla_\mathbf{r}\, n = \boldsymbol{\Gamma}_e$$

Since the electric field is the same for both charged species we have,

$$\boldsymbol{E}_a = \frac{D_i - D_e}{\mu_i + \mu_e} \frac{\nabla_\mathbf{r}\, n}{n}$$

and substituting in $\boldsymbol{\Gamma}_i$ we finally obtain,

$$\boldsymbol{\Gamma}_i = -\left(\frac{\mu_i\, D_e + \mu_e\, D_i}{\mu_i + \mu_e} \right) \nabla_\mathbf{r}\, n = -\, D_a\, \nabla_\mathbf{r}\, n$$

where D_a is called *ambipolar diffusion coefficient*[10]. The last expression states that the flow of massive ions is affected by the mobility μ_e and free diffusion D_e coefficients of electrons through the *ambipolar* electric field $\boldsymbol{E}_a$ calculated above. Since electrons and ions interact through the electric field, their motions are not independent to preserve the plasma quasineutrality.

3.3 Electron and ion waves

The short-range collisions between neutral atoms produce in a gas or liquid zones where particles are compressed together or are spread apart. These local low amplitude pressure and/or density fluctuations produce a sound wave that

[10] The term *ambipolar diffusion* has a different meaning in astrophysics where it refers to the decoupling process of neutral atoms from the plasma during the initial stages of star formation.

propagates in the gas. The situation is more complex in a plasma since it contains ions and electrons that also interact through the plasma electric field. The motion of heavy ions pulls off the lighter electrons to preserve the plasma quasineutrality. Consequently, there are oscillations and waves of plasma charged particles even in the absence of long and short range collisions.

Next, we will derive the expression for the sound velocity in a neutral gas, since this analysis will be useful to study the oscillations of ions and electrons. As we will see, the disparity of the ion and electron movement time scales introduces different plasma waves and oscillations.

We recall first a few concepts[11] before starting, the equation,

$$\boldsymbol{H}(\boldsymbol{r}, t) = \boldsymbol{H}_o \exp[i\, (\boldsymbol{k} \cdot \boldsymbol{r} - \omega\, t)]$$

describes a physical magnitude with *amplitude* $\boldsymbol{H}_o$ periodic in time and space. Along the direction pointed by the *wavenumber vector* $\boldsymbol{k}$ this wave transports energy.

The *wavelength* λ is the distance between two points with equal amplitude (or phase) and $|\boldsymbol{k}| = 2\pi/\lambda$ gives the propagation rate in space. The *period* T is the time elapsed between two points with equal amplitude (or phase) and the *angular frequency* $\omega = 2\pi/T$ characterizes the oscillation in time.

The *phase velocity* $v_f = \lambda/T$ or equivalently $v_f = \omega/|\boldsymbol{k}|$ is the rate at which a point of the wave with constant phase moves along the direction where $\boldsymbol{k}$ points. However, the variation in wave amplitude or *envelope* propagates in space at the *group velocity* $v_g = d\omega/dk$ and $\omega(\boldsymbol{k})$ is called the *dispersion function*. Both speeds v_f and v_g are equal only when $\omega(\boldsymbol{k})$ is a linear function.

3.3.1 Neutral gas sound waves

Neglecting viscosity effects, for the conservation of mass and momentum for ordinary gas of density ρ and velocity $\boldsymbol{u}$ of a neutral gas we have,

$$\frac{\partial \rho}{\partial t} + \nabla_{\mathrm{r}} \cdot (\rho\, \boldsymbol{u}) = 0 \tag{3.38a}$$

$$\rho \left[\frac{\partial \boldsymbol{u}}{\partial t} + (\boldsymbol{u} \cdot \nabla_{\mathrm{r}})\boldsymbol{u} \right] = -\nabla_{\mathrm{r}}\, p_a \tag{3.38b}$$

The equations of motion (3.38) are closed using the thermodynamic relation $p_a\, \rho^{-\gamma} = C$ where C is a constant and $\gamma = C_p/C_v$ is the adiabatic gas constant. This equation expresses that local gas rarefaction and compression are considered as *adiabatic* processes. That is, the molecular collision time scales is very fast and the gas rapidly relaxes to a local equilibrium. We can write,

$$\nabla_{\mathrm{r}}\, p_a = C\, \gamma\, \rho^{(\gamma-1)}\, \nabla_{\mathrm{r}} \rho \quad \text{and then,} \quad \nabla_{\mathrm{r}}\, p_a = \gamma\, p_a\, \frac{\nabla_{\mathrm{r}}\, \rho}{\rho} \tag{3.39}$$

that closes the above equations. We consider the small amplitude oscillations of physical magnitudes ($\boldsymbol{u}$, p_a, ρ) with respect to a fixed equilibrium state of the gas at

[11] See sections 4.1 and 4.2 of reference [6].

rest $\boldsymbol{u}_o \simeq 0$ where gas pressure and density (p_{oa}, ρ_o) are uniform and constant in time. For low amplitude oscillations we can approximate

$$\boldsymbol{u} \simeq \boldsymbol{u}_1 \, e^{i\,(\boldsymbol{k}\cdot\boldsymbol{r}-\omega\, t)}, \quad p_a \simeq p_{ao} + p_{a1}\, e^{i\,(\boldsymbol{k}\cdot\boldsymbol{r}-\omega\, t)} \quad \text{and also,} \quad \rho \simeq \rho_o + \rho_1\, e^{i\,(\boldsymbol{k}\cdot\boldsymbol{r}-\omega\, t)}$$

where ω is the frequency of the oscillation, $\boldsymbol{k}$ is the wave number vector with $|\boldsymbol{k}| = 2\,\pi/\lambda$ where λ is the perturbation wavelength. Neglecting the contribution of second order terms such as $(\boldsymbol{u}_1 \times \rho_1)$, $(p_1 \times \rho_1)$, etc, equations (3.38b) and (3.38a) are linearized. Thus, $(\boldsymbol{u} \cdot \nabla_{\mathbf{r}})\,\boldsymbol{u}$ in equation (3.38b) is of second order and it is neglected, for the pressure gradient (3.39),

$$\nabla_{\mathbf{r}} p_a \sim \gamma\, \frac{p_{ao} + p_{a1}\, e^{i\,(\boldsymbol{k}\cdot\boldsymbol{r}-\omega\, t)}}{\rho_o + \rho_1\, e^{i\,(\boldsymbol{k}\cdot\boldsymbol{r}-\omega\, t)}} \times (i\,\boldsymbol{k})\rho_1\, e^{i\,(\boldsymbol{k}\cdot\boldsymbol{r}-\omega\, t)}$$

and neglecting second order contributions,

$$\nabla_{\mathbf{r}}\, p_a \sim \gamma\,(i\,\boldsymbol{k})\, e^{i\,(-)} \times \frac{\rho_1\, p_o}{\rho_o + \rho_1\, e^{i\,(-)}} \simeq \gamma\,(i\,\boldsymbol{k})\, e^{i\,(-)} \times \frac{p_{ao}\,\rho_1}{\rho_o} \times \underbrace{\frac{1}{1 + e^{i(-)}\,\rho_1/\rho_o}}_{\sim 1}$$

The linearized equations (3.38b) and (3.38a) are,

$$-\omega\,\rho_1 + (\boldsymbol{k}\cdot\boldsymbol{u}_1)\,\rho_o = 0 \tag{3.40}$$

$$-\omega\,\rho_o\,\boldsymbol{u}_1 + \gamma\,\frac{p_o}{\rho_o}\,\rho_1\,\boldsymbol{k} = 0 \tag{3.41}$$

Taking the scalar product with $\boldsymbol{k}$ of equation (3.41) and eliminating the perturbation amplitudes $(\boldsymbol{u}_1 \cdot \boldsymbol{k})$ and ρ_1 is derived the dispersion relation $\omega = c_s\, k$ as,

$$\frac{\omega}{k} = \sqrt{\frac{\gamma\, k_B\, T}{m_a}} \tag{3.42}$$

The pressure and density fluctuations propagate at the sound velocity c_s in a neutral gas and group velocity is equal to phase velocity. The propagation velocity of perturbations depends on the thermal energy and collisions between neutral gas molecules transport pressure waves.

3.3.2 Ion waves

Now we turn our attention to the ion motion that has a slower time scale compared with the electron response. Neglecting collisions, the ion number conservation equation (3.36a) and momentum (3.36b) for ions are,

$$\frac{\partial n_i}{\partial t} + \nabla_{\mathbf{r}} \cdot (n_i\,\boldsymbol{u}_i) = 0 \tag{3.43a}$$

$$m_i\, n_i \left[\frac{\partial \boldsymbol{u}_i}{\partial t} + (\boldsymbol{u}_i \cdot \nabla_{\mathbf{r}})\boldsymbol{u}_i\right] = e\, n_e\, \boldsymbol{E} - \nabla_{\mathbf{r}}\, p_i \tag{3.43b}$$

We will also make use of the *adiabatic approximation* (3.39) $\nabla_r p_i = \gamma (k_B T_i) \nabla_r n_i$ and the field is $\boldsymbol{E} = -\nabla_r \phi$ where ϕ is the plasma electric potential.

The contribution of electrons to the momentum fluctuations can be neglected and its density n_e can be related with the perturbed plasma potential $\phi = \phi_1 \exp[i (\boldsymbol{k} \cdot \boldsymbol{r} - \omega t)]$ using equation (4.6) of volume 1,

$$n_e(\phi) = n_{eo} \exp\left(\frac{e \phi_1}{k_B T_e}\right) = n_{eo}\left[1 + \frac{e \phi_1}{k_B T_e} + \ldots\right] \sim n_{eo} + n_{e1}$$

with $n_{e1} = n_{eo} (e \phi_1 / k_B T_e)$. In the equilibrium state $n_{io} = n_{eo} = n_o$ and $n_{i1} = n_{e1}$ since slow changes in the ion density are rapidly followed by the fast electrons.

As in previous sections 3.3.3 and 3.3.1, equations (3.43) can be linearized and we have,

$$\begin{aligned} -\omega n_{i1} + n_o (\boldsymbol{u}_1 \cdot \boldsymbol{k}) &= 0 \\ -\omega n_{io} m_i \boldsymbol{u}_{i1} + e n_{io} \phi_1 \boldsymbol{k} + (k_B T_i) \gamma n_{i1} \boldsymbol{k} &= 0 \\ n_{i1} - n_o \frac{e \phi_1}{k_B T_e} &= 0 \end{aligned} \tag{3.44}$$

Eliminating the perturbation amplitudes n_{e11}, $(\boldsymbol{k} \cdot \boldsymbol{u}_{i1})$ and ϕ_1 as before,

$$\omega^2 = k^2 \left[\frac{k_B(\gamma T_i + T_e)}{m_i}\right]$$

and the dispersion relation is $\omega = c_{is} k$ where,

$$c_{is} = \sqrt{\frac{k_B(\gamma T_i + T_e)}{m_i}}$$

is the *ion sound speed.* As $c_{is} = \omega / k$ is constant *ion sound* or *ion acoustic waves* propagate without dispersion with identical group and phase wave speeds.

The fluctuations of ions density are similar to sound in neutral gases, the compression and rarefaction of ion density pull the electrons along to shield the fluctuations of the electric field. Ion waves only exist when ions have thermal motion, since c_{is} depends on T_e a high average energy of electrons makes the electric field shielding process faster. The propagation of ion sound waves is independent of ion collisions, contrary to ordinary sound waves in neutral gases.

In PLE plasmas the ion temperature T_i is much lower than $T_e \gg T_i$ thermal energy of electrons is dominant and $c_{is} \simeq \sqrt{k_B T_e / m_i}$. Using the ion plasma frequency $\omega_{pi} = \sqrt{e^2 n_o / m_i \epsilon_o}$ we can write,

$$\frac{\omega^2}{\omega_{pi}^2} \simeq \left(\frac{m_i \epsilon_o}{e^2 n_o}\right) \times \left(\frac{k_B T_e}{m_i}\right) = k^2 \lambda_D^2 = 4\pi^2 \frac{\lambda_D^2}{\lambda^2}$$

where λ represents the wavelength of the perturbation.

The maximum frequency of ion waves is $\omega \sim \omega_{pi}$ corresponding to the shortest perturbation wavelengths $\lambda \sim \lambda_D$ associated with the minimum plasma scale length, in the order of the Debye length (maximum $k = 2\pi/\lambda$). The wave frequency ω decreases for fluctuation wavelengths $\lambda \gg \lambda_D$ (decreasing $k = 2\pi/\lambda$).

3.3.3 Electron waves

We start with the warm plasma model introduced in section 3.2.2. Neglecting collisions, we can drop in equation (3.36) the friction forces, collisional ionization, recombination terms, etc. The continuity (3.36a) and momentum transport (3.36b) for electrons are,

$$\frac{\partial n_e}{\partial t} + \nabla_{\mathbf{r}} \cdot (n_e\, \boldsymbol{u}_e) = 0 \tag{3.45a}$$

$$m_e\, n_e \left[\frac{\partial \boldsymbol{u}_e}{\partial t} + (\boldsymbol{u}_e \cdot \nabla_{\mathbf{r}})\boldsymbol{u}_e\right] = -e\, n_e\, \boldsymbol{E} - \nabla_{\mathbf{r}}\, p_e \tag{3.45b}$$

These are similar to equation (3.38) for a neutral gas except by the plasma electric field $\boldsymbol{E}$ in equation (3.45b). In addition to the collisionless *adiabatic approximation* where $p_e\, n_e^{-\gamma} = cte.$ (see box 3.3) we will need further assumptions involving the electric field to close this system of equations.

First we will consider the fast time scale for the electric field fluctuations that only electrons can follow. In this case, the ions are considered a uniform background fixed in time with uniform density n_{io} and only the electron density $n_e = n_{eo} + n_{e1} \exp(i\, (\boldsymbol{k} \cdot \boldsymbol{r} - \omega\, t))$ fluctuates and $n_{eo} = n_{io} = n_o$ in the state of equilibrium as fast electrons preserve the plasma quasineutrality. Introducing in the Poisson equation $\nabla_{\mathbf{r}} \cdot \boldsymbol{E} = e(n_i - n_e)/\epsilon_o$ the electric field $\boldsymbol{E} = \boldsymbol{E}_1 \exp(i\, (\boldsymbol{k} \cdot \boldsymbol{r} - \omega\, t))$,

$$\nabla_{\mathbf{r}} \cdot \boldsymbol{E} = \frac{e}{\epsilon_o}\left(n_{io} - n_{e1}\, e^{i\,(\boldsymbol{k}\cdot\boldsymbol{r} - \omega\, t)}\right) \quad \text{and then,} \quad i\,(\boldsymbol{k} \cdot \boldsymbol{E}_1) = -\frac{e}{\epsilon_o}\, n_{e1} \tag{3.46}$$

With this linearized equation for the electric field fluctuation amplitude $\boldsymbol{E}_1$ we first explore the limit without thermal fluctuations ($T_e = 0$) setting $\nabla_{\mathbf{r}}\, p_e \sim 0$ in the momentum transport equation (3.45b). Using equation (3.46) and following the steps of section 3.3.1 we obtain from equation (3.45) the linearized equations,

$$\begin{aligned} -\,\omega\, n_{e1} + (\boldsymbol{k} \cdot \boldsymbol{u}_{e1})\, n_o &= 0 \\ -\,i\, m_e\, \omega\, \boldsymbol{u}_{e1} + e\, \boldsymbol{E}_1 &= 0 \\ i\,(\boldsymbol{k} \cdot \boldsymbol{E}_1) + \frac{e}{\epsilon_o}\, n_{e1} &= 0 \end{aligned} \tag{3.47}$$

with the scalar product with $\boldsymbol{k}$ in the second equation and eliminating the perturbation amplitudes n_{e1}, $(\boldsymbol{u}_{e1} \cdot \boldsymbol{k})$ and $\boldsymbol{E}_1$ we find,

$$\omega = \left(\frac{e^2\, n_o}{m_e\, \epsilon_o}\right)^{1/2}$$

We recover the electron plasma frequency introduced in section 4.5.2 of volume 1 as the faster response of electrons to external perturbations and its dispersion relation $\omega(k)$ is independent of the wave number. This means that the disturbance does not propagate in space, the electron low amplitude oscillations are not coupled so far collisions are neglected, as in the derivation of electron plasma frequency ω_{pe} in section 4.5.2 of volume 1.

The next step is to consider the electron temperature $T_e > 0$ using again the *adiabatic approximation* of the warm plasma model. For the pressure gradient $\nabla_\mathbf{r}\, p_e$ in equation (3.45b) we can write[12],

$$p_e = C\, n_e^\gamma, \qquad \frac{\nabla_\mathbf{r}\, p_e}{p_e} = \gamma\, \frac{\nabla_\mathbf{r}\, n_e}{n_e} \quad \text{and,} \quad \nabla_\mathbf{r}\, p_e = \gamma\, (k_B\, T_e)\, \nabla_\mathbf{r}\, n_e \tag{3.48}$$

When the electron pressure fluctuations are *isothermal* we have $\gamma = 1$, whereas $\gamma > 1$ for *adiabatic* transformations[13]. This last equation can be linearized as,

$$\nabla_\mathbf{r}\, p_e \sim \gamma\, (i\, \boldsymbol{k}\, n_1)\, e^{i\, (\boldsymbol{k}\cdot\boldsymbol{r} - \omega\, t)}$$

With this additional term in equation (3.45b) the linearized equations are now,

$$\begin{aligned} -\,\omega\, n_{e1} + (\boldsymbol{k}\cdot\boldsymbol{u}_{e1})\, n_o &= 0 \\ -\,i\, m_e\, n_o \omega\, \boldsymbol{u}_{e1} + e\, n_o\, \boldsymbol{E}_1 + \gamma\, (k_B T_e)\, n_1\, (i\, \boldsymbol{k}) &= 0 \\ i\, (\boldsymbol{k}\cdot\boldsymbol{E}_1) + \frac{e}{\epsilon_o}\, n_{e1} &= 0 \end{aligned} \tag{3.49}$$

Again, the elimination of perturbation amplitudes n_{e1}, $(\boldsymbol{u}_{e1}\cdot\boldsymbol{k})$ and $\boldsymbol{E}_1$ gives,

$$\omega^2 = \frac{e^2\, n_o}{m_e\, \epsilon_o} + \frac{\gamma\, k_B\, T_e}{m_e}\, k^2$$

and introducing the electron thermal velocity $v_{th} = \sqrt{2\, k_B\, T_e/m_e}$ we finally obtain,

$$\omega^2 = \omega_{pe}^2 + \frac{\gamma}{2}\, v_{th}^2\, k^2 \tag{3.50}$$

The phase velocity is $v_f = \omega/k$ and the group velocity v_g can be derived as,

$$2\, \omega\, d\omega = \frac{\gamma}{2}\, v_{th}^2\, (2\, k\, dk) \quad \text{then,} \quad v_g = \frac{\gamma}{2}\, v_{th}^2 \frac{k}{\omega} = \frac{\gamma}{2}\, \frac{v_{th}^2}{v_f}$$

Figure 3.2 shows schematically the relation between the dispersion relation $\omega(k)$ for electron plasma waves and the group and phase velocities. The wavenumber k_s gives the frequency $\omega_s = \omega(k_s)$ in equation (3.50) and $v_f = \omega_s/k_s$ is the slope of the dotted line. The group velocity is different from v_f and $v_g = d\omega/dk$ is the slope of the

[12] Since the ideal gas molecules have a Maxwell–Boltzmann energy distribution. Then, the equation of state $p_e = n_e\, k_B\, T_e$ is consistent with the assumption of a local Maxwellian electron population.

[13] See box 3.3, for the ideal gas law $p_e = k_B T_e\, n_e$ with T_e constant recovered taking $\gamma = 1$ and the adiabatic transformation $p_e = C\, n_e^\gamma$ corresponds to $\gamma > 0$ with C constant.

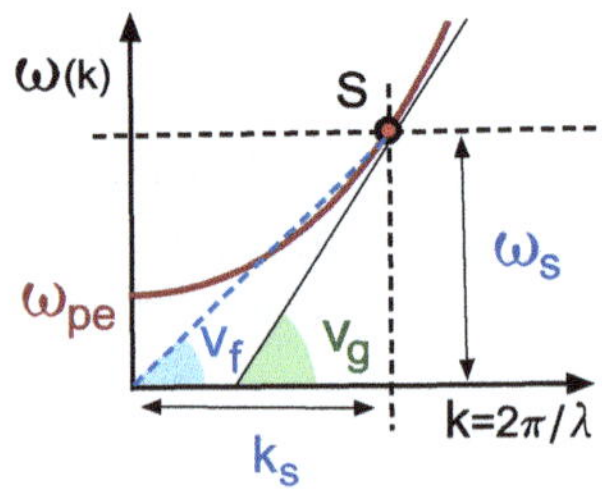

Figure 3.2. Dispersion relation for electron plasma waves.

solid line tangent to the point S. Both, v_g and v_f change as the point S in figure 3.2 moves along the curve (3.50) as $k = 2\pi/\lambda$ increases up to its maximum value for $\lambda \simeq \lambda_D$ corresponding to the minimum plasma scale length for perturbation amplitudes.

Introducing in the dispersion relation (3.50) the electron plasma frequency ω_{pe} we have,

$$\frac{\omega^2}{\omega_{pe}^2} = 1 + \frac{\gamma}{2}\left(\frac{2\, k_B T_e}{m_e}\right) \times \left(\frac{m_e\, \epsilon_o}{e^2\, n_o}\right) k^2 = 1 + \gamma\, \lambda_D^2\, k^2 = 1 + \gamma\, \frac{\lambda_D^2}{\lambda^2}$$

Then, for large perturbation amplitudes λ the frequency of electrons plasma waves is $\omega \simeq \omega_{pe}$ and this oscillation frequency increases as the length scale λ of fluctuations decreases up to a $\omega \simeq \omega_{pe}\sqrt{1+\gamma}$ maximum value for the plasma scale length.

3.4 The Langmuir sheath

Plasmas are limited by solid dielectric or metallic surfaces in nature and in the laboratory. Exposed surfaces of dielectric materials can accumulate electric charge and metallic walls biased at a fixed potential can drain and/or repel ions and electrons from the plasma. Alternatively, charged particles can be also collected by exposed metallic conductors connected to an external electric circuit.

When the electric potential of an exposed metallic surface differs from the distant plasma potential it is adjusted along a characteristic scale length. The steep change in the electric potential profile attached to the wall is called *Langmuir sheath* or *electrostatic plasma sheath.*

The plasma quasineutrality is not accomplished along the plasma sheath and its length scale is crucial for current collection processes from a plasma since the ions (or electrons) are accelerated or repelled by the intense and localized electric field. As figure 3.3 shows schematically, electrons and ions are accelerated or repelled by the intense electric field at the plasma boundaries. The calculation of its spatial potential profile is not trivial, even under the simplest conditions as shows the following one-dimensional model.

The metallic surface of figure 3.3 is negatively biased ($\varphi_w < \varphi_p$) with respect to an unmagnetized and PLE plasma ($T_e \gg T_i$). The ions move along the plasma potential profile $\varphi_p > \varphi(x) > \varphi_w$ and are collected by the wall placed at distance x_w from the

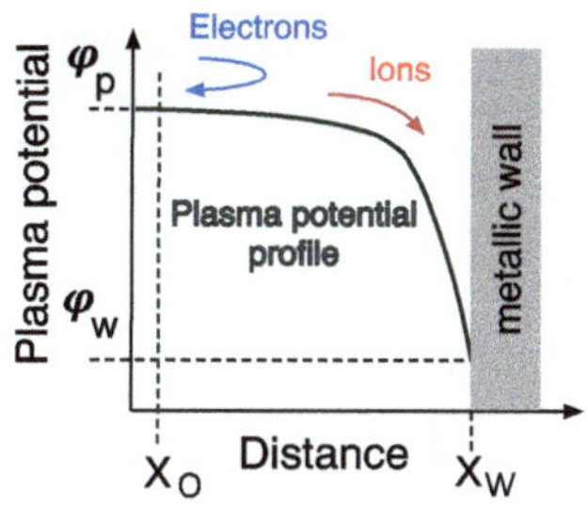

Figure 3.3. Plasma potential profile between the metallic wall and the undisturbed plasma. An intense electric field develops attached to the metallic wall.

unperturbed quasineutral plasma ($n_{eo} \cong n_{io}$) located at $x \leqslant x_o$ where $\varphi_p = \varphi(x)$ is considered uniform.

In low pressure plasmas the mean free path for short-range collisions with neutrals λ_c is usually much longer than the Debye length $\lambda_c \gg \lambda_D$ then, the motion of charges in the region $x_o \leqslant x \leqslant x_w$ of figure 3.3 is collisionless.

We can make use of equations (3.35) in one dimension for these *cold* ions and in this case we can drop the friction term $\boldsymbol{R}_i$ and also $(S_i - L_i) = 0$. In the stationary state we simply have,

$$\frac{\partial}{\partial x}(n_i\, u_i) = 0 \quad \text{and,} \quad m_i\, n_i\, u_i\, \frac{\partial u_i}{\partial x} = +e\, n_i\, E_x$$

where $E_x = -\partial\varphi/\partial x$ and $\varphi(x)$ is the electric potential.

The initial ion speed at x_o is u_{io} and we can write $n_{io}\, u_{io} = n_i(x)\, u_x(x)$ for the ion flux. The second equation can be integrated,

$$\frac{1}{2}m_i u_i^2(x) + e\,(\varphi(x) - \varphi_p) = \frac{1}{2}m_i u_{io}^2$$

so ion energy is preserved in the ion motion. The ion number density can be expressed as,

$$n_i(x) = n_{io}\, u_{io}\left(u_{io}^2 - \frac{2e\,(\varphi - \varphi_p)}{m_i}\right)^{-1/2} \tag{3.51}$$

Since we are interested in the slow time scale of the ion motion we can consider the fast electrons in thermal equilibrium as Maxwellian electron population as in box 3.2.

$$n_e(\varphi) = n_{eo}\exp[-e\,(\varphi - \varphi_p)/k_B T_e] \tag{3.52}$$

The plasma potential profile $\varphi(x)$ can be calculated using the Poisson equation as,

$$\epsilon_o \frac{d^2\varphi}{dx^2} = e\, n_{eo}\left[e^{(e(\varphi-\varphi_p)/k_B T_e)} - u_{io}\left(u_{io}^2 - \frac{2e\,(\varphi - \varphi_p)}{m_i}\right)^{-1/2}\right]$$

The physical magnitudes can be scaled using the ion acoustic speed $c_{ia} = \sqrt{k_B T_e / m_i}$ and,

$$\phi = -e(\varphi - \varphi_p)/K_B T_e > 0 \quad N_e = n_e/n_{eo} \quad N_i = n_i/n_{eo} \quad and, \quad V = u/c_{ia}$$

additionally $2e\varphi/m_i u_{io}^2 = 2\phi/(u_{io}/c_{ia})^2$ and $V_o = u_{io}/c_{ia}$, we obtain the following equation for the dimensionless potential,

$$\frac{d^2\phi}{dx^2} = -\frac{1}{\lambda_D^2}\left[e^{-\phi} - \left(1 + \frac{2\phi}{V_o^2}\right)^{-1/2}\right] \tag{3.53}$$

The distances x and λ_D are intentionally left explicit to show that $s = x/\lambda_D$ defines the length scale of the plasma sheath. The solution $\phi(x)$ of this non-linear differential equation gives the plasma potential profile of figure 3.3 in one dimension. However, no analytical solutions of equation (3.53) have been obtained for $\phi(x)$, however, after some transformations we will find an approximate solution.

3.4.1 Bohm criterion

We introduce $s = x/\lambda_D$ in equation (3.53) where $\mathcal{E} = -d\phi/ds$ represents the scaled electric field, which is normalized to the plasma sheath length scale. Multiplying by $d\phi/ds$ both sides of above equation we have,

$$\int_{so}^{s} \frac{d^2\phi}{ds^2}\frac{d\phi}{ds}ds = \left[\int_{so}^{s}\left(1 + \frac{2\phi}{V_o^2}\right)^{-1/2}\frac{d\phi}{ds}ds - \int_{so}^{s}\exp(-\phi)\frac{d\phi}{ds}ds\right]$$

where $s_o \leqslant s \leqslant s_w$ is the dimensionless coordinate of a point located between s_o and the wall $s_w = x_w/\lambda_D$ in figure 3.3. The integration gives,

$$\frac{1}{2}\left[\mathcal{E}^2(s) - \mathcal{E}^2(s_o)\right] = \frac{1}{2}\left(\frac{d\phi}{ds}\right)^2\Bigg]_{s_o}^{s} = V_o^2\sqrt{1 + \frac{2\phi}{V_o^2}}\Bigg]_{s_o}^{s} + \exp(-\phi)]_{s_o}^{s}$$

The scaled electric field $\mathcal{E}(s) = -d\phi/ds$ is proportional to the slope of the plasma potential profile in figure 3.3 and $\mathcal{E}(s) \gg \mathcal{E}(s_o)$. Then, the term that gives the lower limit of this integral can be neglected since it is small compared with the upper limit term. For points $s > s_o$ within the plasma sheath we have,

$$\frac{1}{2}\mathcal{E}^2(s) = \frac{1}{2}\left(\frac{d\phi}{ds}\right)^2 = V_o^2\left[\sqrt{1 + \frac{2\phi}{V_o^2}} - 1\right] + (\exp(-\phi) - 1) > 0$$

which is always positive. The non-linear differential equation (3.53) for $\phi(x)$ is transformed into a non-linear algebraic equation that can only be solved numerically if we make some approximations. For *ideal plasmas* where $k_B T_e \gg e(\varphi - \varphi_p)$ (or $\phi \ll 1$) the right term can expanded in powers as,

$$\frac{1}{2}\left(\frac{d\phi}{ds}\right)^2 = V_o^2\left[\left(1+\frac{\phi}{V_o^2}-\frac{1}{2}\frac{\phi^2}{V_o^4}+\cdots\right)-1\right]$$
$$\times\left[(1-\phi+\frac{1}{2}\phi^2+\cdots)-1\right]>0$$

and holding the two first powers we have,

$$V_o^2\left[\frac{\phi}{V_o^2}-\frac{1}{2}\frac{\phi^2}{V_o^4}\right]-\phi+\frac{1}{2}\phi^2>0$$

and hence,

$$\left[-\frac{1}{V_o^2}+1\right]\frac{1}{2}\phi^2>0$$

For the last equation to be positive the normalized ion speed has to be $V_o > 1$, that is,

$$V_o > 1, \quad \text{or,} \quad u_{io} > c_{ia} \tag{3.54}$$

This inequality is called *Bohm sheath criterion*, and shows that the speed of ions at point x_o must be *higher* that the ions' acoustic velocity.

We have not derived an *explicit* expression for $\varphi(x)$ but to fulfill this condition an electric field must accelerate the ions to supersonic velocities ($u_{io} > c_{ia}$) *before* the ions enter into the sheath region. Therefore, the plasma potential profile of figure 3.3 is not monotonic and has at least two parts; a *presheath* region before the electron-free zone where the large potential drop of the *plasma sheath* takes place.

Since the electric field is null at s_o, the *presheath* region is located between s_o and the undetermined point s where the condition (3.54) is fulfilled before the large voltage drop associated with the *plasma sheath*. The plasma potential drop along the *presheath* is smooth as,

$$\frac{1}{2}m_i c_{is}^2 = k_B T_e \sim e\,\Delta\varphi \quad \text{then,} \quad \Delta\varphi \sim \frac{k_B T_e}{e}$$

is in the order of the electron thermal energy.

The physical interpretation of the Bohm criterion $u_{io} > c_{ia}$ can be understood with the help of figure 3.4 where the logarithm of electron (3.52) and ion (3.51) plasma densities are represented against the dimensionless potential. The Maxwellian electron number density $n_e = n_{eo}\exp(-\phi)$ is a straight line with slope (-1) and the ion density from equation (3.51),

$$N_i = \left(1+\frac{2\phi}{V_o^2}\right)^{-1/2} \quad \text{and,} \quad \ln(n_i) = \ln(n_{io}) - \frac{1}{2}\ln\left(1+\frac{2\phi}{V_o^2}\right)$$

decreases more slowly with the dimensionless potential. Both start at $\phi = 0$ ($\varphi = \varphi_p$) where $n_{io} = n_{eo}$ and the charge density $\rho = e\,(n_i - n_e) > 0$ must be positive, so n_i

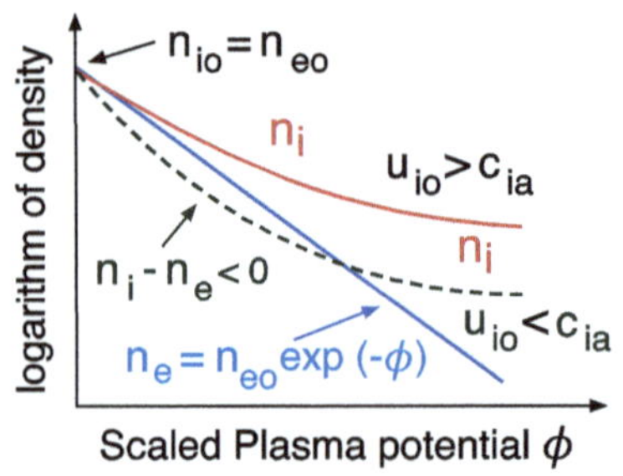

Figure 3.4. The $n_i(\phi)$ and electron $n_e(\phi)$ densities as a function of ϕ.

must always lie over the line corresponding to the electron density. For this to happen the slope of $n_i(\phi)$ at $\phi = 0$,

$$\frac{d}{d\phi}\left[\frac{1}{2}\left(1 + \frac{2\,\phi}{V_o^2}\right)\right] = -\frac{1}{V_o^2} < (-1)$$

needs to be below (-1) and $v_o > 1$, otherwise the electric charge density reverts its sign as in the dotted line of figure 3.4.

So far we have considered a *collisionless* plasma sheath where $\lambda_c \gg \lambda_D$, which governs the thickness of the plasma sheath. When short-range collisions are important the Bohm criterion needs to be generalized and this subject is beyond the scope of the present work.

3.4.2 Child–Langmuir law

The values for $\phi(s)$ are large at points s close to the wall and far from the presheath where the electric potential profile of figure 3.4 becomes steeper. In the Poisson equation (3.53) we can neglect the contribution of electrons $\sim\exp(-\phi)$ in this zone where $2\phi/V_o^2 \gg 1$ and,

$$\frac{d^2\phi}{ds^2} = \left(1 + \frac{2\phi^2}{V_o^2}\right)^{-1/2} \cong \frac{V_o}{\sqrt{2\phi}}$$

Multiplying by $d\phi/ds$ on both sides we can integrate between the point s_i where the electron density is neglected and $s \leqslant s_w$ inside the sheath and close to the wall at $s_w = x_w/\lambda_D$,

$$\int_{s_i}^{s} \frac{d^2\phi}{ds^2}\frac{d\phi}{ds}\,ds = \int_{s_i}^{s} \frac{V_o}{\sqrt{2\phi}}\frac{d\phi}{ds}\,ds \quad \text{then,} \quad \frac{1}{2}\left(\frac{d\phi}{ds}\right)^2\Bigg]_{s_i}^{s} = \sqrt{2}\;V_o\sqrt{\phi}\,\Bigg]_{s_i}^{s}$$

Using the dimensionless electric field $\mathcal{E} = -d\phi/ds$ we have,

$$\frac{1}{2}\left[\mathcal{E}^2(s) - \mathcal{E}^2(s_i)\right] = \sqrt{2}\;V_o\left[\sqrt{\phi(s)} - \sqrt{\phi_i}\right]$$

We neglect $\mathcal{E}^2(s_i)$ compared to $\mathcal{E}^2(s)$ and also $\phi(s) \gg \phi(s_i)$ and the resulting differential equation,

$$\frac{d\phi}{ds} = 2^{3/4}\sqrt{V_o}\,\phi^{1/4}$$

can be integrated for points s close to the wall with electric potential $\phi_w = \phi(s_w)$,

$$\frac{4}{3}\left[\phi_w^{3/4} - \phi^{3/4}\right] = 2^{1/2}\sqrt{V_o}\,(s_w - s)$$

The solution provides an expression for $\phi(s)$ which depends on the ion velocity $V_o = u_{io}/c_{ia}$ and the bias potential of the wall. Neglecting $\phi(s) \ll \phi_w$ we obtain a relation between V_o and ϕ_w as,

$$V_o = \frac{4\sqrt{2}}{9}\frac{\phi_w^{3/2}}{(s_w - s)^2} \quad \text{and substituting,}$$

$$u_{io} = \frac{4\sqrt{2}}{9}\left(\frac{\epsilon_o K_B T_e}{e^2 n_{eo}}\right)\frac{1}{d^2}\sqrt{\frac{K_B T_e}{m_i}}\left(\frac{e\,\varphi_w}{K_B T_e}\right)^{3/2}$$

Since the ion current density is conserved in the sheath we can write $J_i = n_i u_i = n_{io} u_{io}$ and finally we obtain,

$$J_i = \frac{4}{9}\frac{\epsilon_o}{e^2}\sqrt{\frac{2e}{m_i}}\frac{\varphi_w^{3/2}}{d^2} \tag{3.55}$$

which is the *Child–Langmuir law* for a plane diode.

The ion current density J_i from the plasma scales with the 3/2 power of the electric potential φ_w of the wall (or electrode) located at a distance $d = x_w - x$ where x is the point where we consider the potential and the electric field as negligible. This Child–Langmuir law has been accurately verified and also applies to the current of electrons or ions transported between two electrodes separated at distance d.

Similar arguments apply when ions are repelled by the conductive wall of figure 3.3 and replacing m_i by the electron mass m_e in equation (3.55) is also valid for the electron current density.

Turning back to figure 3.3 the potential profile close to the conductive wall can be divided into three zones. Since electrons are repelled, the zone closest to the wall is electron-free and its thickness is given by equation (3.55). Next comes a zone of thickness λ_D a where n_e is appreciable and finally the *presheath* where ions are accelerated up to the Bohm speed (3.54). Its extension is much larger than the *sheath* as the potential drop for ion acceleration along the *presheath* $\Delta\varphi k_B T_e/e$ is smooth.

3.4.3 Space-charge neutralization

The *space charge* refers to the excess of positive or negative electric charge concentrated on a region of space, which produces a localized electric field that rejects charges of the same sign. The maximum current density (positive or negative) that can be transported in vacuum in the steady given by equation (3.55) is said to be *space-charge* limited.

The above Child–Langmuir law gives the maximum ion (or electron) current that can be transported in vacuum under the electric potential drop φ. This is a limiting factor for charged particle sources and specifically for plasma thrusters, since the maximum ion flow that delivers thrust can be bounded by charge space effects

The term *neutralization* refers to mixing positively and negatively charged particles to mitigate or to eliminate these electric fields produced by charged beams. The stream of particles can be *space-charge* neutralized, *current* neutralized or both. In the first case, the mix of positive and negative charges decreases the average beam electric field and the current of opposite charged particle lowers the beam magnetic field in the second case. The result of the neutralization process is a plasma stream of electrons and ions where the quasineutrality condition $n_e \simeq n_i$ holds. Therefore, equation (3.55) is no longer valid as it only accounts for one charged species [7].

As we shall see in chapter 6, electron-emitting cathodes are used in the *space charge neutralization* of ion beams to form plasma thrusters, as the magnetic fields produced are low. This neutralization process creates a quasineutral plasma flow where the ion momentum losses are negligible due to the smallness of the electron mass.

3.5 Commentaries and further reading

The hydrodynamic plasma models accurately describe a large number of physical phenomena despite the gross simplifications made. The derivation of the transport fluid equations essentially comes from reference [5] and an introductory approach is in [2]. An updated and rigorous formulation is found in reference [3] and in the classical book [4] where the calculation of friction force is in chapter 6. The MHD equations are of interest in the study of plasma flows at planetary and galactic length scales and its formulation can be found in the books [2] (section 4.13), [8] (chapter 9) and [9] (chapter 6).

The asymptotic closure of transport equations for neutral gases is discussed in the book [4]. For magnetized plasma see the classical review [5] of Braginskii, and more recently sections 4.10 and 4.11 of reference [2]. The book [10] is a comprehensive and up-to-date formulation of plasma waves theory and experiments.

The study of *plasma sheaths* is of interest in a number of fields. For example, the conductive walls of spacecraft or their exposed metal surfaces can drain electrical charge from the ambient space plasma. In this process called *spacecraft charging* the vehicle or some of the elements can acquire high voltages that produce unwanted currents and/or trigger electric discharges. Reference [11] is an overview of space charging phenomena and [12] a comprehensive book on this subject. Finally, reference [13] is an excellent review on plasma sheaths in the context of space plasmas.

Initially used for the design of vacuum valves, the applications of the Child–Langmuir equation for a plane diode (3.55) have found a countless number of applications. The classical one-dimensional model for a plane sheath is tractable, but the extension of the Child–Langmuir law to more dimensions is difficult, as shown in reference [14]. Additionally, the generalization of Bohm sheath criterion to account for elastic or ionizing collisions in the plasma sheath region is still a subject of current studies.

References

[1] Stone N H 1981 The aerodynamics of bodies in rarefied ionized gas with applications to spacecraft environmental dynamics *Technical Report, NASA Technical Paper* NASA

[2] Fritzpatick R 2015 *Plasma Physics: An Introduction* (Boca Raton, FL: CRC Press)

[3] Woods L C 1993 *An Introduction to the Kinetic Theory of Gases and Magnetoplasmas* (Oxford: Oxford University Press)

[4] Chapman S and Cowling T G 1953 *Mathematical Theory of Non-uniform Gases* (Cambridge: Cambridge University Press)

[5] Braginskii S I 1965 Transport processes in a plasma *Rev. Plasma Phys.* **1** 205–311

[6] Chen F F 1984 *Introduction to Plasma Physics and Controlled Fusion* vol 1 2nd edn (New York: Plenum)

[7] Humphries S 2013 *Charged Particle Beams* 2nd edn (New York: Dove)

[8] Bittencourt J A 2004 *Fundamentals of Plasma Physics* 3rd edn (New York: Springer)

[9] Gurnett D A and Bhattacharjee A 2005 *Introduction to Plasma Physics with Space and Laboratory Applications* (Cambridge: Cambridge University Press)

[10] Stix T 1992 *Waves in Plasmas* (New York: AIP Press-Springer)

[11] Lai S T and Cahoy K 2016 Spacecraft charging *Encyclopedia of Plasma Technology* ed J Leon Shohet (Boca Raton, FL: CRC Press) pp 1352–66

[12] Lai S T 2012 *Fundamentals of Spacecraft Charging: Spacecraft Interactions with Space Plasmas* 4th edn (Princeton, NJ: Princeton University Press)

[13] Robertson S 2013 Sheaths in laboratory and space plasmas *Plasma Phys. Control. Fusion* **55** 093001

[14] Lau Y Y 2001 Simple theory for the two-dimensional Child–Langmuir law *Phys. Rev. Lett.* **87** 278301

IOP Publishing

An Introduction to Plasma Physics and its Space Applications, Volume 2

Basic equations and applications

Luis Conde

Chapter 4

Introduction to electric discharges

> Electric discharges are today frequently employed for low temperature plasma production in the laboratory. They have found a myriad of applications in fields as diverse as microelectronics, surface spectroscopy, biological sterilization of materials, welding, metal cutting, etc. This chapter is a brief introduction to the physical principles of electric discharges and breakdown phenomena.

We are familiar with sparks in electric equipment or with the bright rays in atmospheric storms. Both are examples of *electric breakdown processes* where an avalanche of free charges are produced in liquids, solids or gases by applying an external electric field. For this to occur, the physical medium experiences physicochemical changes that bring to transport electric current in otherwise electrically insulating materials. The term *electric discharge* usually applies to any flow of current through ionized gases and to gas ionization processes produced by the applied electric field. Consequently, the *physics of electric discharges* is a vast multidisciplinary field where plasma physics and material science convene.

The breakdown process in low pressure gases is initiated by a small number of free electrons that are accelerated by the applied electric field up to energies over its ionization threshold. These seeding electrons come from the natural radioactivity or cosmic rays and initiate the *charge avalanche* as they produce electron/ion pairs by the impact ionization process discussed in section 6.2.2 of volume 1. The steady state equilibrium of the electric discharge requires the balance between ionization and charge loss processes introduced in section 6.2.3 of volume 1.

The electric discharge also requires a minimum electric field strength E_b, or equivalently, a *breakdown voltage* V_b established between two electrodes separated

doi:10.1088/978-0-7503-3543-0ch4

by a distance d as in figure 4.1. The low electron impact ionization rate below this threshold does not compensate the electron losses without the action an external ionizing agent (such as UV light, electrons emitted from hot cathodes, etc). This externally sustained charge avalanche is called a *non-self-sustaining discharge*. When the applied voltage exceeds V_d a steady state can be reached and electron–ion LTE or PLE plasmas develop between the electrodes of this *self-sustaining* electric discharge.

The DC gas discharges are sustained by elementary processes at the electrodes, such as the secondary electron emission by ion impact (section 6.4.2 of volume 1), the physical properties of the neutral gas, etc. The breakdown voltages can be also applied by means of electromagnetic waves and the avalanche is less dependent of electrode properties in this case.

Laboratory plasmas are usually produced by electric discharges and in this chapter we will briefly introduce the basic properties of DC and RF electric breakdown phenomena. Historically, electric discharges were studied from the early 19th century using the DC and/or low frequency power supplies available at the time. The ulterior development of high power RF technology through the 20th century led to the observation of electric discharge in radar and microwave communications systems that are now employed in a number of technological applications. The study of light emitted by gas discharges played an important role in the development of modern physics, as its spectra revealed the discontinuous nature of the atomic energy levels and paved the way to quantum mechanics foundations.

4.1 DC electric discharges

The different DC electric discharges in the steady state can be illustrated with the idealized current–voltage (I–V) characteristic curve of figure 4.2, where the discharge voltage V_d against the current I_d of the electrical scheme of figure 4.1(a) is represented. It is valid for pressures below 10 mbar of the gas enclosed in the glass tube.

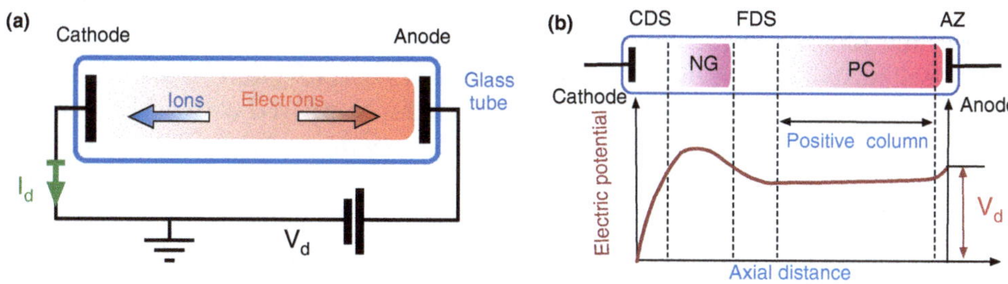

Figure 4.1. (a) Schematic diagram of the DC discharge circuit between two parallel electrodes (see also volume 1, figure 1.3) separated a distance d. (b) The axial variation of the electric potential (solid red line) and the different regions of a low pressure glow discharge. CDS is the cathode dark space, NG negative glow, FDS the Faraday dark space, PC the positive column and AZ the anode zone.

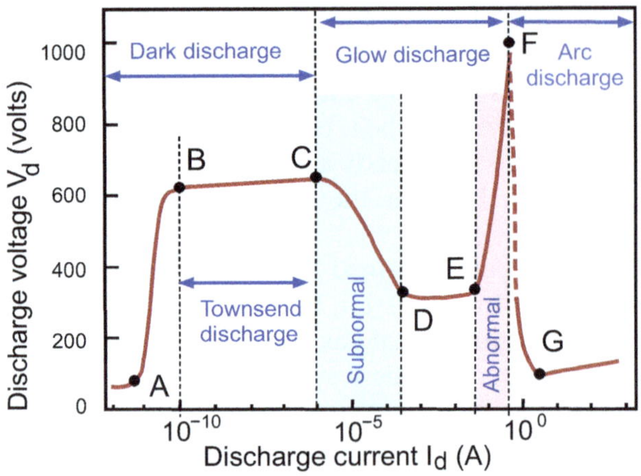

Figure 4.2. The idealized the current–voltage (*I–V*) characteristic curve and the different DC electric discharges regime.

The currents I_d are very low (below 10^{-10}) for applied voltages below the breakdown voltage V_d (between points *A* and *B* in figure 4.2) since the number of charges produced by electrons from cosmic rays or natural radioactivity is small compared with electron losses. For these voltages the discharge is *non-self-sustained* and it would need an additional external electron source to raise the discharge current.

The *self-sustained* discharge triggers when $V_d > V_b$ (point *B*) as the strength of the inter-electrode electric field is high enough and fast electrons can produce the impact ionization of the neutral gas. The electric charge multiplication between *B* and *C* raises by a few orders of magnitude the current I_d in figure 4.2 with small increments in the applied voltage.

Ions are accelerated towards the cathode, whereas electrons move along the opposite direction where they are collected by the anode and lost. However, the impact of energetic ions over the metallic surface of the cathode produces the emission of new secondary electrons (section 6.4.2 of volume 1) that sustain the current through the discharge gap. Additionally, atoms of the cathode material can also be released and this process called *sputtering* is currently employed to deposit thin film layers for optical coatings or semiconductor devices.

This self-sustained *Townsend discharge* between points *B* and *C* is also called *dark discharge* since low visible light is emitted by the gas compared with other regimes in figure 4.2. The currents I_d are still low compared to *glow* and *arc* discharges, so that inelastic collisions of electrons (section 6.2 of volume 1) produce a low amount of light to be observed by the naked eye.

As the discharge current I_d increases further, space charge effects at the electrodes become important and the axial profile of plasma potential becomes non-uniform, as shown in figure 4.1(b). Because of their lower mobility, the ions concentrate in front of the cathode to create a localized intense electric denominated *cathode fall.* Its localized voltage drop is almost equal to V_d and the more mobile electrons proliferate in this region since the high electric field strength increases the local impact ionization rate. Because of this non-uniform voltage profile, higher currents I_d require lower breakdown voltages to sustain the discharge in the *subnormal glow* regime between C and D.

In the *glow discharge* regime (between C and F) the breakdown voltages required to sustain the discharge are lower as the formation of the *cathode fall* is concluded. The discharge current density remains approximately constant and for small increments in V_d the discharge current I_d (in the range 1–100 mA) grows. The reason is the expansion of the discharge volume, so that electric charges are collected over larger surfaces of the electrodes.

The glow discharges can have a complex profile of the axial electric potential, as shown in the scheme of figure 4.1(b). The region close to the cathode where most of the applied voltage V_p drops (typically along few millimeters) is called *cathode dark space* (CDS). Along the adjacent *negative glow* (NG) the plasma potential profile is more smooth and additional regions as the *Faraday dark space* (FDS) and the *positive column* (PC) develop when the inter-electrode distance is large enough. The discharge current flows through the low pressure PLE plasma in the *positive column*, where the velocity of ions and electrons is the sum of a drift component parallel to the electric field and an additional random speed. Finally, electrons are collected at the anode through the small *anode zone* (AZ).

The plateau in the I–V curve ends when the glow discharge covers all the surface of the electrodes, and therefore higher currents I_d require increased discharges voltages in the *abnormal regime* between points E and F. The abrupt transition between F and G to the *arc discharge* regimen takes place for high breakdown voltages and high discharge currents. In arc discharges the current densities are higher and additional physical processes that contribute to the electron multiplication are important. The gas temperature rises by electron–neutral collisions, high electron currents increase the collisional ionization rate, the secondary electron emission at the electrodes becomes important, etc. Consequently, the breakdown voltage drops as the discharge current increases and this ionized gas is essentially an LTE plasma.

4.1.1 Vacuum breakdown in DC fields

The scheme of figure 4.3(a) shows the motion of ions and electrons in the constant electric field $\boldsymbol{E}_d$ established between two parallel electrodes. The drift velocity of electrons given by equation (3.37) is $\boldsymbol{u}_e = \mu_e \boldsymbol{E}_b$ where μ_e the electron mobility and the component proportional to $(\nabla_{\mathbf{r}} n_e / n_e)$ is small and can be neglected in this case.

The number of ionization events produced by one electron in the inter-electrode space can be empirically described using the *first Townsend coefficient*,

$$\alpha = \frac{\langle \nu_i \rangle}{u_d} \tag{4.1}$$

where $\langle \nu_I \rangle$ is the ionization collision frequency introduced in section 6.2.2 of volume 1. Assuming a constant ionization cross section and using equation (2.3) from volume 1

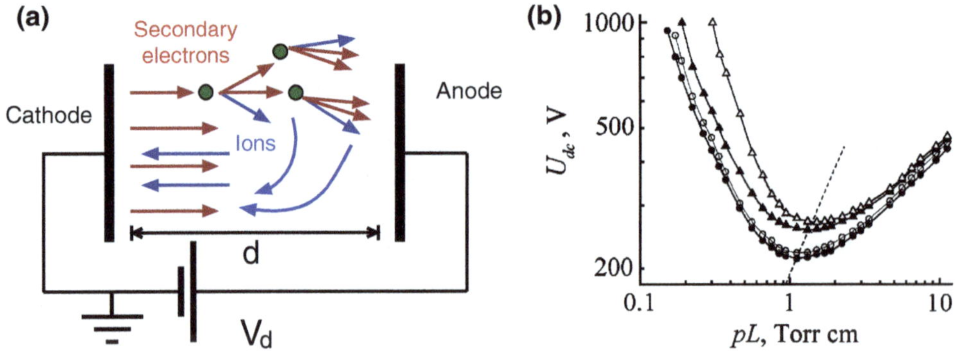

Figure 4.3. (a) Schematic diagram of the exponential electron multiplication in a DC discharge separated a distance *d*. (b) The DC breakdown potential U_{dc} for a glow discharge in argon with different cathode materials. Stainless steel •, platinum ○, nickel ▲, steel △, nickel ■, steel □, platinum ▼ and copper ▽. Reproduced from [1] with permission of IOP. All rights reserved.

we have $\langle \nu_i \rangle \simeq \sigma_I \, n_a \, u_d$ and then $\alpha = 1/\lambda_I$. Therefore, the coefficient α represents the number of ionization mean free paths per unit of length for one electron.

The DC charge avalanche that evolves in both time and space is initiated by a small number of electrons $N_{eo} = N_e(0)$ from the cathode located at $z = 0$. Ionizing collisions produce electron–ion pairs as they move towards the anode and the increment in the electron number along the distance dz is,

$$dN_e(z) = N_{eo} \times (\alpha \, dz)$$

and integration gives,

$$N_e(z) = N_{eo} \exp\left(\int_o^z \alpha(z') \, dz' \right) \simeq N_{eo} \, e^{\alpha \, z} \tag{4.2}$$

This equation where we assume a constant value $\alpha(z) \simeq \alpha$ shows the exponential charge multiplication in the inter-electrode gap. The increment in the electron number between the cathode and anode is,

$$\Delta N_e = N_e(d) - N_{eo} = N_{eo} \, (e^{\alpha \, d} - 1)$$

In the steady state the number of ions impacting the cathode is $\Delta N_i = \Delta N_e$ since ions and electrons are created in pairs. Then, the initial number of electrons $N_{eo} = \gamma_{se} \times \Delta N_i$ equals the secondary emission produced by the ions at the cathode,

$$\gamma_{se} \times \Delta N_i = \gamma_{se} \, N_{eo}(e^{\alpha \, d} - 1) = N_{eo} \quad \text{and,} \quad \gamma_{se} \, (e^{\alpha \, d} - 1) = 1$$

The typical values[1] for γ_{se} in table 4.1 are low and depend on a number of factors, such as the nature of the metal, the state of the surface, etc. Therefore, self-sustained

[1] The values in this table are cited for reference. The secondary electron emission rate depends on a number of factors such as the surface state, cleaning, etc, and different values for γ_{se} can be found in the literature.

Table 4.1. The values for the secondary emission coefficient γ_{se} in equations (4.3) and (4.5) from reference [2]. In this case surface cleaning was made with organic chemicals such as alcohol, methane, etc.

Gas	Cathode	State	γ_{se}
Air	Ni	Cleaned	8.0×10^{-6}–1.5×10^{-4}
N_2	Ni	Cleaned	1.3×10^{-6}–3.7×10^{-4}
N_2	Cu	Cleaned	1.3×10^{-6}
O_2	Cu	Cleaned	$\sim 10^{-7}$
O_2	Cu	Oxidized	$\sim 10^{-6}$
H_2	Cu	Cleaned	$\sim 10^{-6}$
H_2	Cu	Oxidized	5×10^{-5}

Table 4.2. The values of A and B in the empirical Townsend coefficient α (4.4) and their E/P_a range from reference [2]. Gas pressures are in mbar and the electric field in V cm^{-1}.

Gas	A	B	E/p
	cm^{-1} Torr $^{-1}$	V (cm Torr)$^{-1}$	V(cm Torr)$^{-1}$
He	3	34	20–150
Ar	12	180	100–600
Kr	17	240	100–1000
Xe	26	350	200–800
H_2	5	130	150–600
N_2	12	342	100–600
Air	15	365	100–800

cold cathode discharges need a flux of ions and are weakly dependent on the nature of the cathode.

Next, we derive a breakdown condition for a DC self-sustained discharge,

$$\alpha\, d = \ln\left(1 + \frac{1}{\gamma_{se}}\right) \quad \text{or,} \quad \exp\left(\int_o^d \alpha(z')\, dz'\right) = 1 + \frac{1}{\gamma_{se}} \tag{4.3}$$

The coefficient α can be measured in the experiments but its explicit expression is very difficult to calculate. The currently employed empirical formula has been obtained from experimental data fitting.

$$\alpha = A\, p_a \exp\left(-B\, p_a / E_d\right) \tag{4.4}$$

where A and B are characteristic constants for each gas and p_a is the pressure. The empirical values for A and B with their range E/P_a are in table 4.2. Equation (4.4) can be justified as follows; assuming electrons have a Maxwellian distribution, the ionization frequency (equation (6.4) in volume 1) is proportional to,

$$\langle \nu_I \rangle \sim n_a\, u_d\, \exp(-E_I/k_B T_e)$$

where $n_a \sim p_a$ is the neutral atom density and E_I the ionization energy. For an ionization event the electron needs to gain the energy $\Delta E = E_p \times \lambda_I$ from the DC electric along the distance λ_I of one ionization mean free path. Thus,

$$\frac{E_I}{k_B T_e} \sim \frac{E_I}{E_b\, \lambda_I} = \frac{E_I \times n_a \times \sigma_I}{E_b} \sim \frac{B}{(E_d/p_a)}$$

Using equation (4.1) we finally have,

$$\alpha = \frac{\langle \nu_i \rangle}{u_d} \sim A\, p_a\, \exp\left(-\frac{B}{(E_d/p_a)}\right)$$

Finally, for the breakdown voltage $V_b = E_b \times d$ combining equation (4.3) with equation (4.4) we have,

$$\frac{1}{d} \ln\left(1 + \frac{1}{\gamma_{se}}\right) = \alpha = A\, p_a\, \exp\left(-\frac{B\,(p_a\, d)}{V_b}\right)$$

and we obtain,

$$V_b = \frac{B\,(p_a\, d)}{\ln[A\,(p_a\, d)] - \ln\,[\ln(1 + 1/\gamma_{se})]} \tag{4.5}$$

This expression is called *Paschen law* and gives DC breakdown voltage V_b between parallel electrodes, that essentially depends on the product $(p_a\, d)$ and the secondary electron emission coefficient γ_{se} of the cathode.

The breakdown voltage of equation (4.5) is represented in figure 4.3(b) for different gases as a function of $(p_a\, d)$. The curves have a minimum at,

$$(p_a\, d)_{\min} = \frac{e}{A} \ln\left(1 + \frac{1}{\gamma_{se}}\right)$$

where $(V_b)_{\min} = B\,(p_a\, d)_{\min}$ and for $(p_a\, d) > (p_a\, d)_{\min}$ increases roughly linearly as the electric field decreases when d grows.

Figure 4.3(b) shows the values of breakdown voltages observed in the experiments are in agreement with the predictions of this simple model for cathodes of different conductive materials. The voltage V_b scales with the product $(p_a\, d)$ and for large values of $(p_a\, d)$, it increases essentially linearly, whereas for low $(p_a\, d)$ there is a limiting value. For a fixed pressure p_a, the energy gained by electrons decreases with the inter-electrode distance and therefore the ionization rate decreases and higher V_d are required to trigger the discharge. In contrast, the electrons have less ionizing collisions before reaching the anode when, d is small, so the ion production also decreases when $(p_a\ d) < (p_a\ d)_{min}$. The minimum breakdown voltage in figure 4.3(b) gives the optimum condition for the electron avalanche.

4.2 Microwave discharges

The material of electrodes in a DC discharge is conductive since the electric current I_d is transported from cathode to anode as shown in the scheme of figure 4.3. Their surfaces are subjected to the bombardment of high energy ions and electrons that can alter their surface properties by the sputtering process, etc. The eventual emission of material particles from the electrodes has deleterious effects for some practical applications such as spectrochemical analysis, thin film deposition, etc. This problem can be mitigated in part by applying an alternating voltage between the electrodes that will alternatively act as cathode and anode.

The breakdown under AC electric fields depends on its frequency and applied voltage provided all other parameters (gas, electrode separation, etc) are fixed, so an additional time scale is introduced. The motion between the electrodes determines the slow time scale where processes that occur are similar to DC voltages and this low frequency regime is called *mobility controlled* (as in section 4.1.1). For typical frequencies below 1 MHz the Paschen law in equation (4.5) is still valid, as long as the time scale of applied RF voltage is slow enough. In order for the charges to reach the electrodes, the maximum frequency for the mobility controlled regime depends on practical details such as gas pressure and device size.

At higher electric field frequencies, the ions oscillate under the fast electric field in the discharge region but the more mobile electrons can reach the electrodes. Increasing the frequency further, (typically over 10 MHz) there is also a minimum time scale for the electron motion and they also oscillate. The ionization of the neutral gas increases by these confined electrons and this charge multiplication process, typically in the microwave frequency range, is denominated *diffusion controlled* breakdown.

Therefore, plasma breakdown in high frequency electric fields is different from the DC case since the electron avalanche may only develop in time at high frequencies, but not in space. The typical frequencies for RF waves are in the range from 1 kHz to 1 GHz, most commercial equipment operates at 13.56 MHz as is the standard frequency allotted by the international radio authorities. The *microwave discharges* are produced by electromagnetic fields with wavelengths in the order of centimeters (within the 1–300 GHz range) used by communication equipment. In RF discharges the plasma is dominated by the displacement currents and two main variants exist;

- In the *capacitive coupled discharges* the plasma is confined between two parallel electrodes. One is usually grounded and the second is connected to an RF generator which applies a several hundred volts amplitude signal.
- In the *inductively coupled discharges* the plasma is coupled by means of an alternating magnetic field. The coil is connected to the RF generator and the sinusoidal magnetic field induces an AC electric field in the plasma volume

acting as the secondary coil of a transformer. The primary external coils can be made of a winding around a glass or quartz tube or antenna.

The time evolution of the number of charges is given by the continuity equation (3.19),

$$\frac{\partial n_e}{\partial t} + \nabla_{\mathbf{r}} \cdot (n_e \, \boldsymbol{u}_e) = S_e - L_e$$

Here the source term $S_e = \nu_I \, n_e$ is related to the ionization frequency ν_I introduced in section 6.2.2 of volume 1 and $S_e = -\beta_r \, n_e^2$ is the electron losses by recombination (6.5). In the discharge equilibrium $\partial n_e / \partial t = 0$ and using equation (3.37),

$$D_e \nabla_{\mathbf{r}}^2 \, n_e \, + \nu_I \, n_e - \beta_r \, n_e^2 = 0$$

then, the solutions of this non-linear partial differential equation give the $n_e(\boldsymbol{r})$ density profile in the inter-electrode gap. This equation can be linearized for low pressures since the losses by volume recombination are usually small; the typical dimension of the discharge region is much shorter than the recombination mean free path and we have,

$$D_e \nabla_{\mathbf{r}}^2 \, n_e \, + \nu_I \, n_e = 0$$

Electrons are produced by the impact ionization of neutrals with frequency ν_I and are lost from the discharge bulk as they diffuse.

So far we considered D_e uniform and in this simplified model the electron diffusion coefficient D_e, the ionization ν_I frequency, recombination coefficient β_r, etc, are different for all gases and also a function of the electric field strength, pressure. They have to be determined from a thorough combination of experiments and theoretical calculations.

4.3 Breakdown in high frequency fields

In the oscillatory electric field $\boldsymbol{E}(t) = \boldsymbol{E}_o \, e^{i \, \omega_o \, t}$ with frequency $\omega_o = 2\pi \, f_o$ the equilibrium results from the balance between the elastic collisions of electrons and the energy gained from the field. The equation for the component of the electron velocity $\boldsymbol{u}_e(t)$ in the direction parallel to the field can be expressed as,

$$m_e \, \dot{u}_e = -e \, E(t) - m_e \, \nu_e \, u_e$$

where ν_m is the collision frequency for momentum transfer. Assuming the electron motion as $x(t) = x_o \, e^{i \, \omega_o \, t}$ and solving for x_o its velocity in the steady state is,

$$u_e(t) = \frac{e \, E_o}{m_e} \frac{(i \, \omega_o - \nu_m)}{\omega_o^2 + \nu_m^2} \, e^{i \, \omega_o \, t}$$

The electron current density is $j = -e \, n_e \, u_e$ and averaging the product of real parts of J and $E(t)$ over the period $2\pi / \omega_o$ we derive the input power,

$$\langle P \rangle = \frac{e^2\, E_o^2\, \nu_m}{2\, m_e(\omega_o^2 + \nu_m^2)}$$

that compensates the energy losses.

The energy loss term of electrons by elastic collisions with neutrals of mass m_a is,

$$\frac{m_e}{m_a}\, n_e\, \nu_m\, \langle W \rangle$$

where m_e/m_a is the fraction of energy transferred to the atom in each encounter, ν_m the collision frequency and $\langle W \rangle$ is the average energy of one electron. The ionization energy is E_I and $n_e\, \nu_I\, E_I$ represents the average energy consumed in the production of new particles with ionization collision frequency ν_I. Finally, for the characteristic diffusion time scale τ_d the energy transported by the diffusion of electrons is, $n_e\langle W \rangle/\tau_d$.

The average energy balance then reads,

$$\langle P \rangle = \frac{e^2\, E_o^2\, \nu_m}{2\, m_e(\omega_o^2 + \nu_m^2)} = \frac{m_a}{m_e}\, n_e\, \nu_m\, \langle W \rangle + \frac{n_e}{\tau_d}\langle W \rangle + n_e\, \nu_I\, E_I \tag{4.6}$$

When diffusion and ionization terms can be neglected compared with the elastic collisions we can derive,

$$\langle W \rangle = \frac{m_a}{m_e^2}\, \frac{e^2\, E_o^2\, \nu_m}{2\, m_e(\omega_o^2 + \nu_m^2)} = E_{k,m}\, \frac{m_a}{m_e}\, \frac{\omega_o^2}{(\omega_o^2 + \nu_m^2)}$$

Here $E_{k,m} = e^2\, E_o^2/(2\, m_e\, \omega_o^2)$ is the peak kinetic energy $m_e\, u_{e,m}^2/2$ the electron reach under the electric field. The electron energy increments with the mass ratio $(m_a/m_e) \gg 1$ since the energy transference from electrons to the neutral gas atoms by elastic collisions is low.

Solving equation (4.6) for E_o we derive an *breakdown condition* for a RF plasma and we have two limits and the different regimes are schematically shown in figure 4.4(a). For high gas pressures where $\nu_m \gg \omega_o$ electrons can reach a high average energy $\langle W \rangle \sim E_I$ as losses to the walls are small and elastic collisions are dominant. In this limit equation (4.6) becomes,

$$E_o = \left[\frac{m_e^2}{m_a\, e^2}\, E_I\, (\omega_o^2 + \nu_m^2)\right]^{1/2}$$

The strength of the electric field increases approximately linearly with p_a as the elastic collisions with neutrals increase with the gas pressure.

In the opposite limit of low pressures $\nu_m \ll \omega_o$ elastic collisions are neglected in equation (4.6) and,

$$E_o = \left[\frac{2\, m_e\, (\omega_o^2 + \nu_m^2)}{e^2\, \nu_m}\left(\langle \nu_I \rangle\, E_I + \frac{\langle W \rangle}{\tau_d}\right)\right]^{1/2}$$

The electric field breakdown E_o rises as the elastic collision frequency ν_m decreases with the gas pressure.

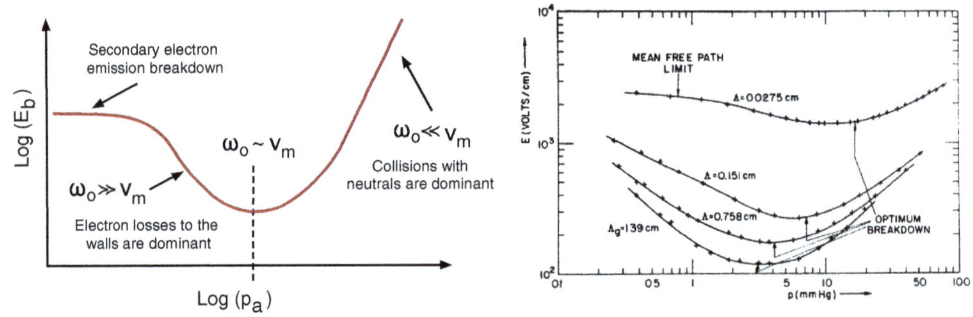

Figure 4.4. The diagram of the microwave breakdown electric fiend strength E_b as a function of the gas pressure p_a (left) and the curves for hydrogen for 10 cm wavelength (right). Reproduced from [3], copyright 1949, with permission of APS.

Then, the general shape of the breakdown curves (also called RF Paschen curves) is shown in figure 4.4 and is similar to the DC Paschen curves but using the gas pressure p_a instead of the product (p_a d). For high pressures elastic collisions are dominant and the electric field strength increases for pressures lower than the minimum where $\omega_o \sim \nu_m$ since more ionizing collisions are needed.

However, the electric field threshold for charge multiplication does not increase indefinitely as the pressure decreases, since an additional charge production becomes important. Electrons do not experience ionizing collisions for very low pressures but their impacts on the surfaces of electrodes can produce additional electrons by the secondary emission. These in turn create new negative charges, etc, this material-dependent electron multiplication process is the *multipactor discharge* that we will study in the next chapter.

4.4 Commentaries and further reading

Electric discharges are the common source of laboratory plasmas and this fact motivates this introductory chapter. The interested reader can find a much more complete and systematic discussion in the monographs [2, 4] and [5]. Recent reviews on this subject are [6] and [7] and microwave and RF breakdown are introduced in the classical book [8].

The books [2] and [5] contain a compilation of relevant data of practical interest, such as diffusion coefficients, Townsend coefficient α, the values of E/P, etc, required for theoretical calculations and experiments as well. The fundamentals of electric discharges and their practical applications are introduced in the monograph [9].

References

[1] Lisovskiy V, Yakovin S and Yegorenkov V 2000 Low-pressure gas breakdown in uniform dc electric field *J. Phys. D: Appl. Phys.* **33** 2722–30

[2] Raizer Y P 1991 *Gas Discharge Physics* (Berlin: Springer)

[3] Brown S C and MacDonald A 1949 Limits for the diffusion theory of high frequency gas discharge breakdown *Phys. Rev.* **76** 1629–33

[4] Lieberman M A and Lichtenberg A J 1994 *Principles of Plasma Discharges and Materials Processing* (New York: Wiley)
[5] Brown S C 1994 *Basic Data of Plasma Physics. The Fundamental Data on Electrical Discharges in Gases* 2nd edn (New York: American Vacuum Society Classics, AIP Press)
[6] Braithwaite N S J 2000 Introduction to gas discharges *Plasma Sour. Sci. Technol.* **9** 517–27
[7] von Kudell A and Schulz-von der Gathen A 2017 Foundations of low-temperature plasma physics-and introduction *Plasma Sour. Sci. Technol.* **26** 11301
[8] McDaniel A D 1966 *Microwave Breakdown in Gases* (New York: Wiley)
[9] Harry J E 2010 *Introduction to Plasma Technology* 1st edn (Weinheim: Wiley)

IOP Publishing

Chapter 5

The multipactor discharge

The *multipactor* is an electron multiplication process harmful for RF equipment of spacecraft. The intense electric field accelerates electrons to high energies inside the communication components and collisions with the materials at the walls can produce secondary electron emission. This material-dependent electron multiplication in vacuum is called *multipactor discharge*. In this chapter are discussed the basic characteristics of this electric breakdown and the calculations its discharge thresholds using a classical model.

In RF equipment and/or microwave components, such as filters, multiplexers, etc, the electromagnetic field is bounded by metallic or dielectric walls and usually disposed within small volumes. The electron impact ionization frequency decreases with the gas pressure and figure 4.4 shows the electric field strength for RF and microwave breakdown increases for gas pressures below threshold value.

When the gas pressure is low enough that the mean free path $\lambda_c \sim 1/(n_a\,\sigma_c) \ll L_s$ for the characteristic electron–atom collisions is below the characteristic size L_s of the system (or equivalently $\nu_{ea} \sim 0$) electrons can impact the exposed surfaces. In this case, the secondary electron emission from the walls discussed in section 6.4.3 of volume 1 can produce new electrons and, under certain circumstances, an explosive charge multiplication process called *multipactor discharge*. Then, the electric field strength for breakdown decreases in figure 4.4 since the other physical mechanism for charge production takes over at higher pressures.

The multipactor charge avalanche is often undesired as it dissipates an important fraction of the input power and may eventually lead to the catastrophic failure of the

doi:10.1088/978-0-7503-3543-0ch5

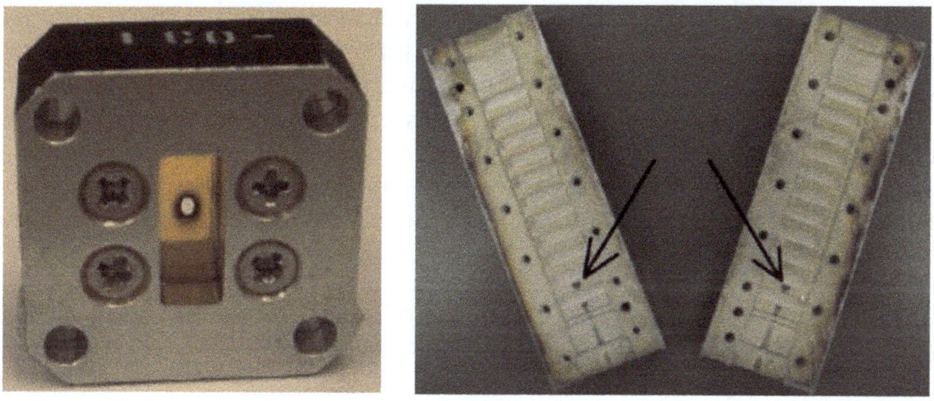

Figure 5.1. Two examples of the harmful effect of multipactor electron discharge in microwave components (reproduced with permission of ESA). The left picture shows the hole produced by this discharge in a Kapton window. The arrows in the right photograph indicate the depletion on the surface of two halves of an aluminum microwave low pass filter produced by the multipactor discharge. The release of heat also blackened the sides of the equipment in this case.

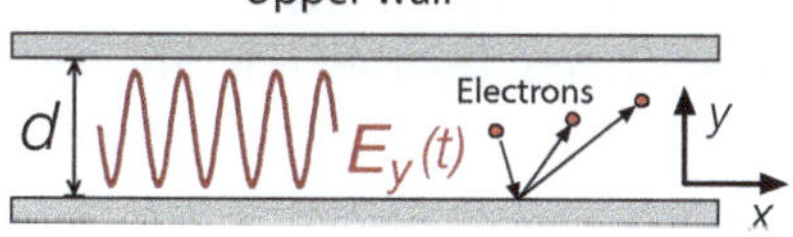

Figure 5.2. Schematic diagram of the sorter section d of an idealized component and the electric field $E_y(t)$.

system. The photographs in figure 5.1 illustrate the harmful effects of the electron multipactor in two microwave components tested in vacuum.

The high electric currents caused by the multiplication of negative charges produced the hole in the window of a Kapton[1] dielectric material in the left picture. The right photograph shows two halves of an aluminum low pass filter and the arrows indicate the black spot created on their inner surfaces by uncontrolled stray currents. The released heat has also blackened both sides of this equipment.

Consequently, the prediction of the radio frequency RF and microwave electric breakdown thresholds is of paramount importance in fields as diverse as satellite telecommunications, particle accelerators or semiconductor processing systems. In this chapter we will introduce the basic characteristics of the *electron multipactor* and the classical model used to predict the discharge thresholds.

5.1 The electron multipactor discharge

The RF or microwave equipment can be roughly modeled in one dimension as in figure 5.2, where the distance $L_s \sim d$ represents the characteristic length of the

[1] Kapton is a high performance plastic material employed in a number of applications in the aerospace industry.

system, corresponding to its smaller section. The sinusoidal electric field $E_y = E_o \sin(\omega_o t)$ of the electromagnetic wave $E_y(t) = V(t)/d$ of frequency $\omega_o = 2\pi f_o$ is parallel to the Y axis and propagates along the X direction.

The figure 5.3(a) shows the measurements of breakdown voltages $V_o = E_o \times d$ between two parallel stainless steel electrodes (1.0 cm in diameter) separated $d = 55$ millimeters as a function of the argon pressure [1]. The three main breakdown regions are represented in the scheme of figure 5.3(b) for clarity. In this experiment the electrical discharges were produced by an electromagnetic wave of frequency f_o= 50 MHz as in the scheme of figure 5.2.

The *electron multipactor* lies below $p_a \sim 10^{-4}$ mbar and the denominated *multipactor plasma* between this pressure and 2×10^{-2} mbar. Finally, the RF Paschen[2] curve is for pressures over 0.02 mbar.

The figure 5.4 shows schematically the charge production processes in the three breakdown regimes. Since $\lambda_c \sim 1/p_a$ (equivalently $\nu_c \sim p_a$) the ratio λ_c/d defines the different RF charge multiplication regimes. When $1 \gg \lambda_c/d$ the electron impact ionization of gas dominates and for $1 \ll \lambda_c/d$ secondary electron emission controls the charge avalanche. The intermediate regime $\lambda_c \sim d$ where both charge production mechanisms coexist is the *collisional multipactor* or *multipactor plasma* regime.

In the electron multipactor the electron–atom collision frequency $\nu_{ea} \sim 0$ and the motion of negative charges driven by the RF electric field inside the cavity is collisionless. Only the impact at the walls of high energy electrons can produce additional negative charges by secondary emission. Hence, the *multipactor* breakdown essentially relies on the physical properties of the material. The secondary electron emission is responsible for charge multiplication and the maximum and minimum values of V_o in figure 5.3 are related with the two crossover energies E_1 and E_2 of the SEY curve of figure 6.13(a) of volume 1, as the electron emission takes place within a range of electron energies.

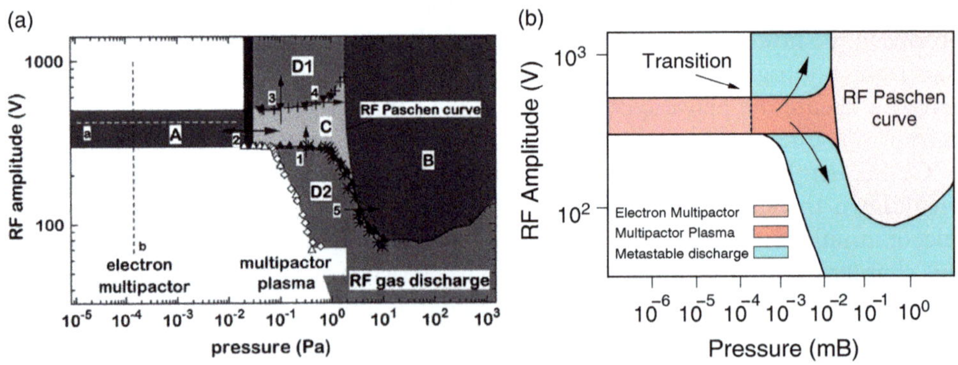

Figure 5.3. The left figure shows actual experimental data curves of RF breakdown peak voltages as a function of the neutral gas pressure from reference [1] and is reproduced with permission of AIP. These zones are schematically represented in the right figure as shadowed regions for clarity.

[2] Represented as a function of the pressure p_a not the product $p_s \times d$.

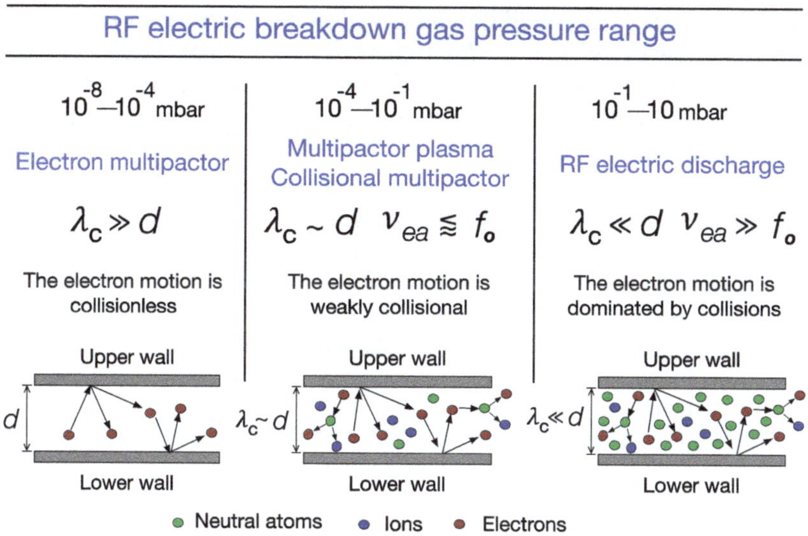

Figure 5.4. The scheme of the different RF electric breakdown regimes as a function of the neutral gas pressure inside the cavity of figure 5.2. The frequency of the RF wave is f_o, the dominant collisional mean free path is $\lambda_c = 1/(\sigma_c\, n_a)$ and ν_{ea} the collision frequency between electrons and neutral atoms.

The opposite limit where breakdown is controlled by electron impact ionization of gas atoms and was discussed in section 4.2. Since the collision frequency $\nu_{ea} \sim n_a \sim p_a$ is proportional to the gas pressure, the RF wave frequencies $f_o = \omega_o/2\pi$ are typically $\nu_{ea} \gg f_o$. Ions and electrons result from the impact ionization of neutral gas atoms, whereas the contribution of secondary electron emission from the walls is negligible. This electric breakdown is strongly dependent on the properties of the neutral gas filling the cavity of figure 5.2. For low RF frequencies, the breakdown regime is diffusion controlled (see section 4.2), peak discharge voltages can be characterized the RF Paschen curves introduced in section 4.1.1, as shown in figures 4.4 and 5.3.

Finally, in the intermediate *multipactor plasma* regime both charge production mechanisms coexist and the weight of each depends on the range of neutral gas pressures as shown in the scheme of figure 5.4. As both characteristic lengths of the problem are comparable, the theoretical models are difficult.

5.2 The multipactor discharge model

The electric power thresholds for the multipactor breakdown regime can be predicted using a simple one-dimensional model valid for the RF frequency range. For low gas pressures the electron collisions with the ambient atoms can be neglected [2, 3]. In this case, figure 5.4 shows the main physical mechanism for charge multiplication is the secondary electron emission from metallic or dielectric surfaces. The impact of electrons moving between the two walls driven by the oscillating electric field $E_y(t)$ as in the scheme of figure 5.2 can produce the negative charge avalanche under certain conditions.

The distance d has to be small enough to allow the electron to transverse the gap of figure 5.2 in a few cycles $T_o = 2\,\pi/\omega_o$ (where $\omega_o = 2\,\pi\,f_o$) of the electric field. As we shall see, the stable charge avalanche also requires *resonant orbits* for electrons where secondaries are emitted at a certain the phase with respect to the RF field.

In the one-dimensional scheme of figure 5.2 one electron moves along the Y axis subjected to a sinusoidal electric field $E(t) = E_o \sin(\omega_o t)$ and no force is applied along the X direction. In the following we will only consider its motion along the direction parallel to the electric field as the force exerted by the RF magnetic field can be neglected. The electron equation of motion is,

$$m_e \frac{du_e}{dt} = -e\,E_o \sin(\omega_o\,t)$$

and direct integration gives,

$$u_e(t) = A - \frac{e\,E_o}{m_e\,\omega_o} \cos(\omega_o\,t)$$

$$y_e(t) = B + A\,t + \frac{e\,E_o}{m_e\,\omega_o^2} \sin(\omega_o\,t)$$

where constants A and B are determined by the initial conditions. The secondary electron is emitted at the time t_e from the lower plate located at $y_e(t_e) = 0$ with speed $u_o > 0$ and we have,

$$u_e(t) = \left[u_o + \frac{e\,E_o}{m_e\,\omega_o} \cos(\omega_o\,t_e)\right] - \left(\frac{e\,E_o}{m_e\,\omega_o}\right) \cos(\omega_o\,t) \tag{5.1}$$

This velocity can be seen as the sum of a an oscillatory component proportional to $\cos(\omega_o\,t)$ plus a constant drift speed,

$$u_{ed} = u_o + \frac{e\,E_o}{m_e\,\omega_o} \cos(\omega_o\,t_e) > 0$$

where the electron emission velocity u_o is a material-dependent parameter. The incident electron has to impact the surface at $y = d$ in figure 5.2 at its maximum speed u_f for the efficient production of secondaries. Setting in equation (5.1) $\alpha = \omega_o\,t_e$ and $\omega_o\,t = n\,\pi + \alpha$, where n is an odd integer,

$$u_f = u_o + 2\left(\frac{eE_o}{m_e\omega_o}\right)\cos(\alpha) > 0 \tag{5.2}$$

Then, the acceleration of the released electron to velocities $u_f > u_o$ by the RF electric field requires the phase α in the range $-\pi/2 \leqslant \alpha \leqslant +\pi/2$.

The electron position $0 \leqslant y_e(t) \leqslant d$ is given by,

$$\begin{aligned} y_e(t) = u_o\,(t - t_e) + \left[u_o + \frac{e\,E_o}{m_e\,\omega_o} \cos(\omega_o\,t_e)\right](t - t_e) \\ + \frac{e\,E_o}{m_e\,\omega_o^2}[\sin(\omega_o\,t) - \sin(\omega_o\,t_e)] \end{aligned} \tag{5.3}$$

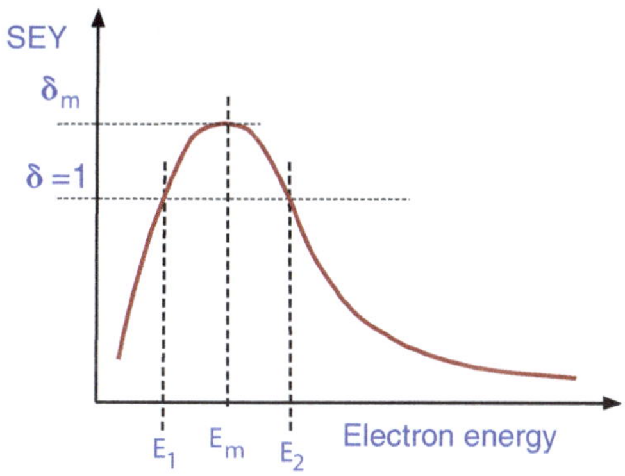

Figure 5.5. The typical SEY curve $\delta(\epsilon)$ represented against the electron energy.

The electron multiplication requires impact on the opposite surface located at $y = d$ in the next half-period of the wave to produce new secondaries. We can use again $\alpha = \omega_o\, t_e$ and $\omega_o\, t = n\,\pi + \alpha$ in equations (5.1) and (5.3) and with,

$$\sin\,(\omega_o\, t) = \sin\,(n\,\pi + \alpha) = -\sin\,(\alpha)$$

equation (5.3) becomes,

$$d = \frac{u_o}{\omega_o} + \frac{e\,E_o}{m_e\,\omega_o^2}[2\,\sin\,(\alpha) + (n\,\pi)\cos\,(\alpha)]$$

Here $E_o = V_{rf}/d$ is the amplitude of the sinusoidal electric field required for the electron avalanche and finally we obtain,

$$V_{rf} = \frac{m_e}{e}\,(\omega_o\, d)\frac{d\,\omega_o - u_o\,(n\,\pi)}{2\,\sin\,(\alpha) + (n\,\pi)\cos\,(\alpha)} \tag{5.4}$$

The RF voltage amplitude $V_{rf} = E_o/d$ is a function of the initial phase α, an odd integer n and the product $(d\,\omega_o)$.

Then, breakdown voltages V_{rf} in equation (5.4) rely on the undetermined phase α and the secondary electron emission velocity u_o. The former depends on the time t_e when the electron is emitted with respect to the RF electric field and the latter relies on the material properties. Since we have little information of both parameters we are forced to introduce some simplifications at this point.

In first place we will simplify the physical description of the secondary electron emission process from metallic or dielectric surfaces, that was previously introduced in section 6.4.3 of volume 1. It depends on physical magnitudes such as the kinetic energy E of impacting electron, its angle of incidence, etc. However, it can be roughly approximated for perpendicular impact by the secondary emission yield (SEY) coefficient curve of figure 5.5. It can be characterized only by its two *crossover energies* (E_1, E_2) where $\delta(E_{1,2}) = 1$ and its maximum SEY value $\delta_m > 1$ at the energy E_m. We will make use only of these few parameters that can be experimentally determined, and table 5.1 shows their typical values for some materials.

Table 5.1. Characteristic values of the SEY curve of figure 5.5 for different materials from table A.6 in reference [4]. The energies are in eV and δ_m is a dimensionless parameter.

Gas	δ_m	E_1	E_2	E_m
Gold	1.79	150	4000	1000
Silver	2.22	30	5000	165
Aluminum	2.98	30	5000	805
Alodine	1.83	41	5000	180
Copper	2.25	25	5000	175

The secondary electron emission of electrons is observed within the $E_1 \leqslant E \leqslant E_2$ energy range where $\delta(E) > 1$ and the multipactor breakdown thresholds of figure 5.3 reflect this fact. For gas pressures below $p_a \sim 10^{-4}$ mbar the electron multiplication takes place between a maximum and a minimum V_{rf} voltage of the RF signal. These bounds are related to the crossover energies E_1 and E_2 of figure 5.5 as they give the minimum and maximum electron velocities where the electron multiplication can be produced.

The next simplification is the assumption of a constant ratio $k = u_f/u_o$ between the impact velocity u_f and the emission velocity u_o within the $E_1 \leqslant E \leqslant E_2$ energy range[3] and using equation (5.2) we obtain,

$$u_f = \frac{k}{k-1}\left(\frac{e}{m_e}\right)\frac{V_{rf}}{(d\ \omega_o)}\ 2\cos(\alpha) \tag{5.5}$$

Eliminating $u_o = u_f/k$ from equation (5.5) we find,

$$V_{rf} = \left(\frac{m_e}{e}\right)\frac{(d\ \omega_o)^2}{2\sin(\alpha) + (n\ \pi)\left[\dfrac{k-1}{k+1}\right]\cos(\alpha)} \tag{5.6}$$

Its minimum value for fixed $(d\ \omega_o)$ corresponds to the phase α_m which maximizes,

$$\phi_n(\alpha) = 2\sin(\alpha) + (n\ \pi)\left[\frac{k-1}{k+1}\right]\cos(\alpha)$$

and $(d\phi_n/d\alpha) = 0$ gives,

$$\tan(\alpha_m) = \left[\frac{k-1}{k+1}\right]\frac{2}{n\ \pi} \tag{5.7}$$

where n is an odd integer.

Combining equations (5.5) and (5.6) the breakdown voltage V_{rf} can be eliminated,

[3] As we shall, see the value of k can be estimated by the experimental data fitting of breakdown voltages.

$$u_f \frac{k-1}{k} \frac{1}{2\cos(\alpha)} = \frac{(d\ \omega_o)}{\phi_n(\alpha)}$$

and $(d\ \omega_o)$ is related to the electron impact velocity $u_f = \sqrt{2\ e\ E/m_e}$ as,

$$(d\ \omega_o) = \sqrt{\frac{2\ e\ E}{m_e}} \left(\frac{k-1}{k}\right)\left[\tan(\alpha) + \left(\frac{n\ \pi}{2}\right)\frac{k+1}{k-1}\right] > 0 \qquad (5.8)$$

For the charge multiplication to occur, the electron kinetic energy must lie within $E_1 \leqslant E \leqslant E_2$ of figure 5.5 and equation (5.8) defines the range, $(d \times \omega_o)_1 \leqslant (d \times \omega_o) \leqslant (d \times \omega_o)_2$ where the breakdown can be triggered holding α and n fixed. Additionally, for $(d\ \omega_o) > 0$ since $\tan(\alpha) < 0$ for $-\pi/2 \leqslant \alpha \leqslant 0$ for constant u_f and k in equation (5.8) a minimum phase $\alpha_l < 0$ (see equation 5.2) is required.

Equations (5.6)–(5.8) suggest the multipactor breakdown can be analyzed using the *susceptibility charts* where the RF amplitude V_{rf} is represented against $(d\ \omega_o)$. Despite the crude simplifications made, the RF breakdown voltage thresholds observed in the experiments follow the basic predictions of this simple model. Specifically, equation (5.8) predicts a minimum value for $(d\ \omega_o)$ and $V_{rf} \sim (d\ \omega_o)^2$ for constant phase within the range $\alpha_l \leqslant \alpha \leqslant \alpha_m$ given by equations (5.8) and (5.7).

In figure 5.6 the experimental values of V_{rf} have been represented against the product of $(d \times f_o)$ which has a minimum value. The maximum and minimum RF breakdown voltages are approximately disposed along two straight lines of slope two in logarithmic scales. For fixed $(d\omega_o)$, the experimental data over the lower branch corresponds to the minimum voltage V_{rf} predicted by equation (5.6) for the constant phase α_m given by equation (5.8), which maximize $\phi_n(\alpha)$ in equation (5.6). The measurements of V_{rf} disposed along the upper branch correspond to the α_l minimum phase. Therefore, the experimental data fitting would permit fixing the free parameters α_l and k in equations (5.6)–(5.8).

The susceptibility curves predicted by the model that can be compared with the experimental data of figure 5.6 are shown in figure 5.7 also in logarithmic scales. On the left is drawn the closed area corresponding to the first mode $n = 1$ and for $n = 1$, 3, 5, 7, 9 are on the right graph under the same conditions. First, using values of reference [2] we set $k = 2.5$ and $\alpha_l = -64.7°$ that can be adjusted later to fit with the available experimental data. The lower straight line is depicted using equation (5.6) for the constant α_m calculated with equation (5.7) and the upper line corresponds to the fixed α_l minimum phase.

Next, multipaction requires electron impact energies $E_1 \leqslant E \leqslant E_2$ where $\delta(E) > 1$ in figure 5.5 and equation (5.8) gives $d\ \omega_o$ for $\alpha_l \leqslant \alpha \leqslant \alpha_m$ holding the electron impact energy E constant. Using the two crossover energies of table 5.1 for the specific material and equations (5.8) and (5.6) the upper and lower curves that close the enclosure of figure 5.7 are calculated. This process can be repeated for all odd integers n as shown by the graph in figure 5.7(b) and the model predicts the RF electron breakdown for all points within the closed areas. The minimum RF breakdown voltage V_{rf} can be estimated by the dashed line depicted in figure 5.7

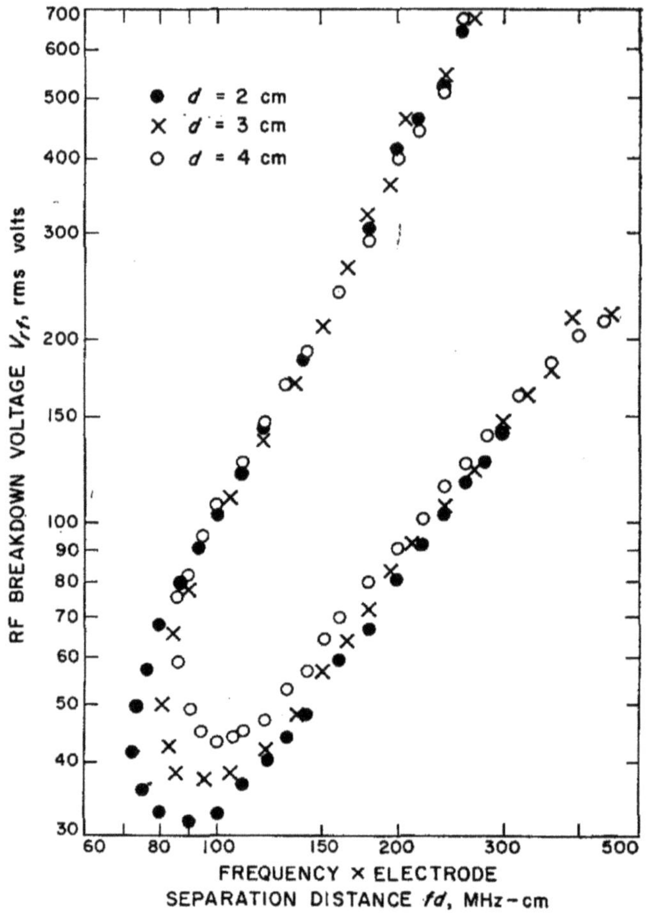

Figure 5.6. The multipactor breakdown voltages between coaxial electrodes of radii a and $b > a$ separated a distance d. Reproduced from reference [5] with permission from AIP.

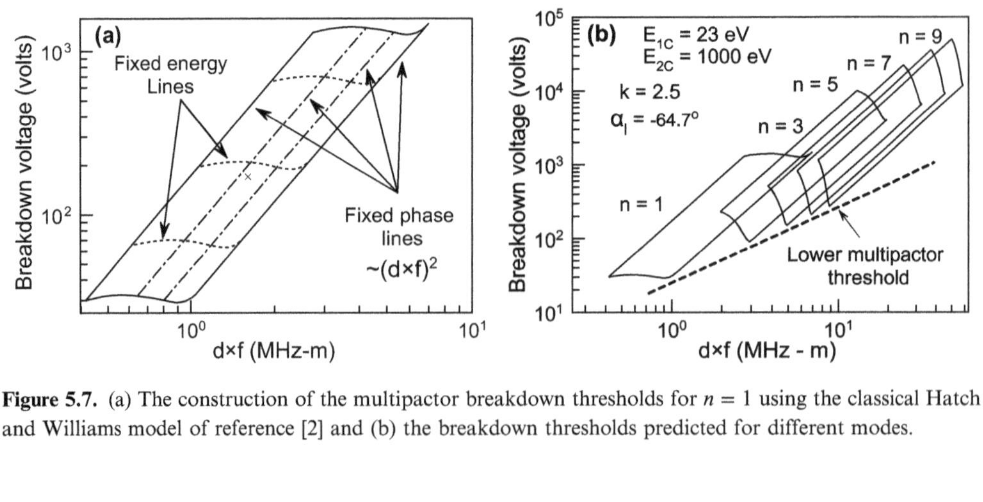

Figure 5.7. (a) The construction of the multipactor breakdown thresholds for $n = 1$ using the classical Hatch and Williams model of reference [2] and (b) the breakdown thresholds predicted for different modes.

and the values of the constant k and α_l that fit the experimental data can be also obtained using those of the upper and lower branches of figure 5.6.

5.3 Commentaries and further reading

Strictly speaking, the model for the multipactor discharge discussed in section is valid for a limited time scale. The electric field in the inter-electrode space is altered by the electron charge avalanche and their motion is not only driven by the external electric field when space charge effects becomes important. However, the susceptibility graphs obtained are in reasonable agreement with the experimental observations.

The basic model of reference [2] has been extended to more involved physical situations. These include different geometries as coaxial electrodes [5, 6], dielectric materials at the walls [7] or *single surface* multipactor [8]. This breakdown develops when the distance d in figure 5.2 is large enough so the electron returns to the emitting surface of emission before impacting the other wall.

The main practical application of the model discussed above is the prediction of multipactor discharge in space-based communication equipment and RF accelerators of charged particles. The practical testing of multipactor breakdown in RF components for space communication systems is thoroughly discussed in references [4] and [9].

The numerical estimations of breakdown thresholds save expensive laboratory tests of RF components and are very useful for their design. The gas discharge breakdown process in connection with high power microwave filter design is discussed in [10], and reference [11] reviews the multipactor discharge theories in a historical context.

The principles introduced in section 5.2 are also on the basis of computer simulations of multipactor discharge, These are based on individual particle tracking schemes [12, 13], particle-in-cell (PIC) [14] or Monte Carlo codes (see [7, 8] and the references therein). The ESA multipactor calculator software can be downloaded from a public access website [15] under request. Commercial software for microwave component design as *FEST*$_3$*D* [16], the *SPARK*$_3$*D* [17] or *CST Studio Suite* [18] also include multipactor and/or gas discharge simulation modules.

However, the reliability of numerical simulations critically depends on the values of the SEY coefficient. Unfortunately, it strongly relies on factors hard to control, such as the nature of the material, the physical state of its surfaces, the preparation of the samples, etc. Table 5.1 gives only a few approximate values, however, more accurate measurements for many materials under a wide variety of experimental conditions can be found in the scientific literature.

In the computer simulations, the SEY curve of figure 5.5 is usually numerically approximated by analytical expressions, as the Waugham's empirical formula [3, 19]. The SEY is calculated on the basis of the three energies of table 5.1 and these calculations are essentially valid for electron emission perpendicular to the emitting wall and low impact energies. This simple expression has been recently improved in reference [20].

Finally, for the intermediate pressure range of 10^{-4}–10^{-2} mbar in figure 5.3 the collision frequency ν_{ea} between electrons and neutral gas atoms and RF wave frequency f_{RF} are similar, or of the same order of magnitude. In this case two charge production mechanisms coexist and electrons are produced by either, the impact ionization of neutrals and by the secondary emission at the walls, as shown in figure 5.4. The physical description of this *collisional multipactor* requires more involved models as in references [13, 21–23] to consider the elastic and inelastic electron collisions.

References

[1] Höhn W J, Beckmann R and Wilhem R 1997 The transition of a multipactor to a low-pressure gas discharge *Phys. Plasmas* **4** 940–4

[2] Hatch A and Williams H B 1954 Multipacting modes of high frequency gaseous breakdown *Phys. Rev. Lett.* **112** 681–5

[3] Vaugham J R M 1989 A new formula for secondary emission yield *IEEE Trans. Plasma Sci.* **36** 1963–7

[4] 2003 Multipaction design and test. ECSS-E-20-01A *Technical Report* European Cooperation for Space Standardization

[5] Woo R 1968 Multipacting discharges between coaxial cylinders *J. Appl. Phys.* **39** 1528–33

[6] Kossyi I, Luk'yanchikov G, Semenov V, Zhanova N, Anderson D, Lisak and Puech J 2010 Experimental and numerical investigation of multipactor discharges in a coaxial waveguide *J. Phys. D: Appl. Phys.* **43** 345206

[7] Kishek R and Lau Y 1998 Multipactor discharge on a dielectric *Phys. Rev. Lett.* **80** 193–6

[8] Iqbal A, Verboncoeur J and Zhang P 2019 Temporal multiparticle Monte Carlo simulation of dual frequency single surface multipactor *Phys. Plasmas* **26** 024503

[9] 2013 Multipaction design and test. ECSS-E-20-01A Rev.1 *Technical Report* European Cooperation for Space Standardization

[10] Kudsia C, Boria V and Cogollos S 2018 High-power considerations in microwave filter networks *Microwave Filters for Communication Systems: Fundamentals, Design, and Applications* ed R Cameron, C Kudsia and R Mansour (Chichester: Wiley) ch 20 pp 643–77

[11] Kishek R 1998 Multipactor discharge on metals and dielectrics: Historical review and recent theories *Phys. Plasmas* **5** 2120–6

[12] Pérez F, Conde L, Alfonseca M, Galan L and Raboso D 2008 CEST and MEST: Tools for the simulation of radio frequency electric discharges in waveguides *Simul. Model. Pract. Theory* **16** 1438–52

[13] Conde L, Pérez F, del Lara J, Alfonseca M and Raboso D 2009 Langevin simulation of RF collisional multipactor breakdown of gases *Phys. Rev. E* **79** 066403

[14] Gopinath V, Verboncoeur J and Birdsall C 1998 Multipactor electron discharge physics using an improved secondary emission model *Phys. Plasmas* **5** 1535–40

[15] ESA multipactor software tool http://multipactor.esa.int/

[16] FEST3D Microwave filter design software http://fest3d.com/

[17] SPARK3D Multipactor and corona analysis http://www.3ds.com/

[18] CST Studio Suite https://www.3ds.com/

[19] Vaugham J R M 1989 Multipactor IEEE Trans *Electron Dev* **35** 1172–80

[20] Ludwick J, Tripathi G, Cahay M, Fairchild S, Murray P and Back T 2018 A new fit to secondary emission yield in the low impact voltage regime: an improvement of Vaughan's expression *AIP Adv.* **8** 085017

[21] Riyopoulos S 2004 Collisional multipactor inside ambient gas *Phys. Plasmas* **11** 2036–45

[22] Udiljak R, Anderson D, Lisak M and Puech J 2004 Improved model for multipactor in low pressure gas *Phys. Plasmas* **11** 5022–31

[23] Kim H and Verboncoeur J 2006 Transition window breakdown from vacuum multipactor discharge to RF plasma *Phys. Plasmas* **13** 123506

IOP Publishing

An Introduction to Plasma Physics and its Space Applications, Volume 2

Basic equations and applications

Luis Conde

Chapter 6

Introduction to plasma propulsion in space

Plasma thrusters are nowadays one of the more relevant applications of plasma physics in space technology and are used for in-space propulsion. The electric power is employed to accelerate a plasma stream or the propellant mass up to exhaust velocities one or two orders of magnitude faster than those delivered by conventional chemical propulsion. This performance drastically reduces the amount of fuel needed for a given maneuver, allowing longer mission times and heavier payloads. In this chapter we discuss the basics of electric propulsion in space and introduce some relevant plasma thruster technologies.

Plasma thrusters are used for in-space propulsion of spacecraft due to a combination of practical and economic reasons. The electric power is employed in these propulsive systems to accelerate the propellant up to exhaust velocities one or two orders of magnitude faster than those achieved in conventional chemical propulsion. Hence, the Tsiolkovsky or rocket equation (see section 6.2 and figure 6.2) shows the amount of propellant necessary for satellite maneuvers is consequently reduced. The optimal use of on-board fuel available is critical since it determines relevant issues such as the satellite mass, its size, etc. Therefore, significant propellant weight savings allow longer mission times and heavier payloads.

The rationale is to foster the economic competitiveness of satellite systems since additional services can be made available using heavier payloads. Additionally, electric thrusters are today the only propulsion technology available for some deep space missions. Today, hundreds of different electric thrusters have been tested and successfully used for orbit corrections as well as in deep-space scientific missions [1].

6.1 Basic orbital maneuvers

Nowadays, the main users of plasma thrusters commercially available are the large telecommunication satellites operated in a circular geostationary orbit (GEO; orbit altitude 35 786 km). The amount of propellant that can be made available on board is critical as they need orbit transfers and/or maneuvers along the long operational

doi:10.1088/978-0-7503-3543-0ch6

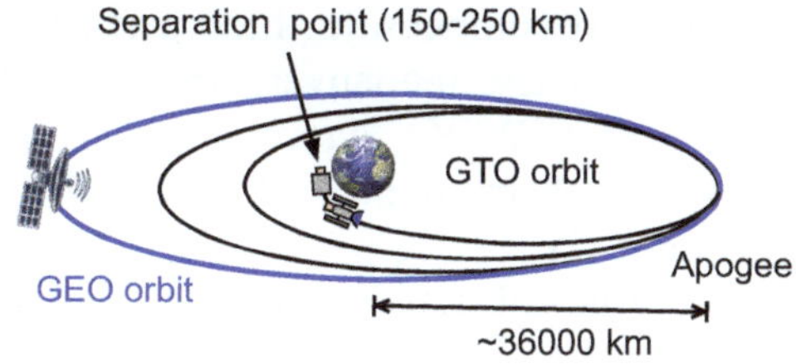

Figure 6.1. Schematic representation of the orbit transfer process of a geostationary satellite.

Table 6.1. Typical values of Δv in equation (6.3) for different maneuvers and orbit transfers from reference [3].

Maneuver	Δv (m s^{-1}) per year
Drag compensation	30
Station keeping GEO	50
Orbit transfer	Δv (km s^{-1})
Earth-LEO	9.7
LEO-GEO	4.3
GTO-GEO	1.5
Interplanetary	Δv (km s^{-1})
Nearby planets	5–8
Far planets	10–15

life of these satellites. The equatorial GEO orbit is reached by a sequence of different propellant-consuming orbital maneuvers and is schematically shown in figure 6.1, where the satellite speed $\Delta v = v_f - v_i$ changes when the thruster is fired. Reference values of Δv for orbit transfers and satellite maneuvers are in table 6.1.

In first place, these satellites describe an elliptical geosynchronous transfer orbit (GTO) after its separation from the third stage of the launch rocket. To move the satellite to a quasi-circular orbit on the Earth's equatorial plane, a thrust of about 400–500 N is applied at its apogee that increases in $\Delta v = 1.5$ km s^{-1} its speed. This transfer process from GTO to GEO orbit where the satellite motion plane with respect to the equator is also corrected, usually takes time and several thruster firings. This basic scheme is modified by a number of specific details, such as the latitude of the launch site, the maximum thrust delivered by the launcher rocket, the inclination of its initial GTO orbit with respect to the Earth's equatorial plane, etc.

The orbital drag caused by the high altitude atmospheric gases is negligible at GEO orbit. However, small perturbations in the orbital motion of satellites are produced by the lunar and solar gravitational interactions and/or the effects of the irregular distribution of Earth's mass. Hence, satellite thrusters are fired on a daily basis to deliver 80–100 mN to compensate these perturbations during the 15–20 years duration of its mission. These maneuvers are currently denominated north–south (NSSK) and east–west (EWSK) *station keeping* and are crucial for telecommunication satellites since their uplink and downlink high gain antennas must point along precise directions [2].

Eventual collisions with other objects in orbit also require additional propellant consumption and more thrust is applied at the end of the satellite operational lifetime, since there are a limited number of slots (~400 satellites) available in GEO orbit. To avoid its eventual return to Earth's atmosphere, the satellite velocity is usually incremented in $\Delta v = 3.0 \text{ m s}^{-1}$ to raise its altitude about 300 km up into a graveyard orbit. This maneuver is intended to mitigate the proliferation of debris at lower altitudes [2].

Similarly, the reduced propellant available on-board also limits the operational life of GPS systems, operating in medium Earth orbits (MEO, altitudes from below 36 000 km down to 2000 km). Their precise and synchronized motions also require periodic thruster firings for orbit corrections as well as for the end-of-life satellite retirement.

Furthermore, new constellations of interconnected small satellites (100–700 kg) intended for interactive television and planetary Internet coverage are being deployed in low Earth orbit (LEO, altitudes from 160 to 2000 km). Their long-term interaction with the high altitude atmosphere is not negligible and they need to be continuously operated to compensate the orbital drag force. The size and weight of these satellites are an issue for the stowage of fuel required for orbital maneuvers and also the final satellite disposal. Conventional chemical thrusters are not an option in this case and electric propulsion is the only available technology so far.

We can conclude that service time of expensive satellites in Earth's orbit is directly related with the amount of fuel that can be stowed on board. The acceleration of charged particles by electric fields is an attractive idea, as the Tsiolkovsky equation shows that important reductions can achieved increasing the exhaust velocity of the propellant. Plasma thrusters make use of different physical mechanisms to accelerate the ejected flow of particles up to higher speeds than conventional chemical propulsion.

6.2 The Tsiolkovsky equation

The ejected propellant mass provides thrust to the spacecraft and the Tsiolkovsky or rocket equation describes the relationship between its velocity and the mass of the system. Momentum conservation (see chapter 2 of reference [4]) gives the thrust that is delivered by the mass of propellant $m_p(t)$ ejected with constant speed v_{ex} with respect to the spacecraft frame of reference as,

$$F = -\frac{d}{dt}(m_p\, v_{ex}) = -v_{ex}\,\frac{dm_p}{dt} = -v_{ex}\,\dot{m}_p$$

where $\dot{m}_p$ is also constant in time. The total mass of the vehicle is $m(t) = m_p(t) + m_f$ the sum of the propellant plus the final delivered mass m_f in the maneuver which is fixed.

Since $\dot{m}_p = m_{po}/\tau_b$ where τ_b is the burning time at a constant rate, we have $\dot{m} = \dot{m}_p$ and the equation of motion for the spacecraft with speed $v(t)$ is,

$$m\,\frac{dv}{dt} = -\,v_{ex}\,\frac{dm}{dt} \quad \text{then,} \quad \int_{v_{si}}^{v_f} dv = -v_{ex}\int_{m_f+m_{po}}^{m_f}\frac{dm}{m}$$

Here $m_o = m_f + m_{po}$ is the initial mass of the spacecraft and writing $\Delta v = v_f - v_i$ we have,

$$\frac{\Delta v}{v_{ex}} = \ln\left(\frac{m_{po}+m_f}{m_f}\right) \text{ or equivalently, } \Delta v = I_{sp}\,g_o\ln\left(1+\frac{m_{po}}{m_f}\right) \tag{6.1}$$

The last equation relates the increment Δv of satellite speed as required for the maneuver with the *specific impulse* $I_{sp} = v_{ex}/g_o$ where g_o is the standard Earth's gravity.

This last magnitude can also be understood as the ratio between the thrust delivered F and the weight of consumed propellant per time unit as,

$$I_{sp} = \frac{F}{\dot{m}_p\,g_o}$$

and is used to compare the performance of thrusters. Equation (6.1) can be cast as,

$$\exp\left(\frac{\Delta v}{v_{ex}}\right) = \frac{m_o}{m_f} = \frac{m_o}{m_o - m_{po}} \tag{6.2}$$

The decreasing ratio between the initial mass of the spacecraft $m_o = m_f + m_{po}$ and the mass of propellant m_{po} required for the increment Δv is,

$$\frac{m_{po}}{m_o} = 1 - e^{-\Delta v/v_{ex}} \tag{6.3}$$

and is represented in figure 6.2(a) as a function of the exhaust propellant speed.

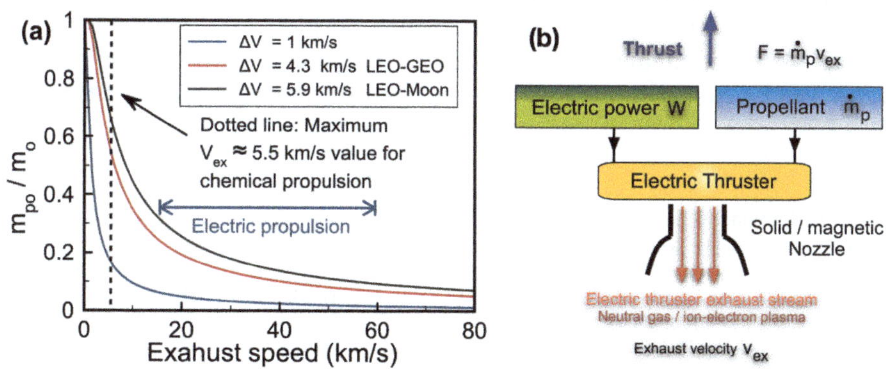

Figure 6.2. (a) The calculations of the ratio m_{po}/m_o as a function of the exhaust propellant velocity v_{ex} using the Tsiolkovsky equation (6.3) for typical values of Δv of table 6.1. The basic scheme of an electric thruster is in (b) where $\dot{m}_p$ is the propellant mass flow rate with exhaust speed v_{ex} and W is the electric power consumption.

6.3 Thrust and specific impulse

Rocket engines operate according to the principle of action and reaction where the gas exhaust speed v_{ex} is crucial for momentum transfer. The thrust is essentially delivered by the *momentum thrust* term $F = \dot{m}_p\, v_{ex}$ as the low ambient pressures in space make the *pressure thrust* negligible [4]. Under ideal conditions equation (6.1) relates the ratio m_{po}/m_f between the mass of fuel m_{po} needed to increase in $\Delta v = v_f - v_i$ the velocity of a satellite with final mass m_f (dry mass) as,

$$\frac{m_{po}}{m_f} = e^{\Delta v/v_{ex}} - 1 \tag{6.4}$$

The integration of the thrust force $F(t)$ over the burning time t_b gives the *total impulse*,

$$I_t = \int_o^{\tau_b} F(t)\, dt$$

and for constant thrust reduces to $I_t = F \times \tau_b$. The *specific impulse* I_{sp} is the *total impulse per unit weight* of propellant as,

$$I_{sp} = \frac{\int_o^{\tau_b} F(t)\, dt}{g_o \int_o^{\tau_b} \dot{m}_p(t)\, dt} = \frac{I_t}{m_p\, g_o}$$

where g_o is the standard Earth's gravity. For constant thrust F and mass flow $\dot{m}_p$,

$$I_{sp} = \frac{I_t}{m_p\, g_o} = \frac{F \times \tau_b}{\dot{m}_p \times \tau_b \times g_o} = \frac{F}{\dot{m}_p\, g_o}$$

and using $F = \dot{m}_p\, v_{ex}$ we have also $I_{sp} = v_{ex}/g_o$. Then, I_{sp} represents the *thrust delivered by the weight of propellant consumed by time unit* and measures the efficiency of the propulsive system.

Table 6.2 shows the typical values of thrust F, burning time τ_b and specific impulse for different conventional chemical thrusters. In figure 6.2(a) is represented the ratio m_{po}/m_o given by equation (6.3) against the exhaust speed v_{ex} for different v values of table 6.1.

Table 6.2. Performance data and standard propellants employed by conventional chemical thrusters from reference [3]. The typical burning time τ_b is indicated in minutes (m), hours (h) or seconds (s).

Thruster type	I_{sp}(s)	F(N)	τ_b(s)	Fuel
Cold gas	50	0.1–100	~m	N_2, ammonia
Liquid	200	1–500	~h	Hydrazine
Liquid	300–400	10^6	~m	$LO_2 + H_2$, kerosene
Solid	250	10^7	100 s	Powder (Al, Mg, Zn, S)

Cold nitrogen or ammonia gases ejected at low speeds (~0.49 km s^{-1}) are employed for low thrust maneuvers. Higher yields (up to 10^6–10^7) are delivered by mixtures of reactive liquids or solid chemicals producing hot gases with typical velocities in the range 2–4 km s^{-1}. For conventional chemical propulsion systems (such as those of table 6.2) the maximum propellant exhaust velocity that can be achieved is $v_{ex} \lesssim 5.5$ km s^{-1}, since it is limited by the energy produced by chemical reactions [4].

This upper bound for v_{ex} is indicated in figure 6.2(a) by the vertical dotted line and the ratio m_{po}/m_o decreases dramatically as the propellant exhaust velocity is increased. For example, the Space Shuttle main engines deliver hot gases with $v_{ex} \sim 4.4$ km s^{-1} and for a GTO-GEO transfer ($\Delta v = 1.5$ km s^{-1}) we obtain $m_{po}/m_o = 0.405$ which gives $m_{po} = 405$ kg for a 1000 kg satellite. Increasing $v_{ex}\sim 20$ km s^{-1} this calculation gives $m_{po}/m_o = 0.078$ reducing to $m_p = 78$ kg the weight of fuel.

The specific impulse in table 6.2 characterizes the efficiency of each propulsive system. Propellants based on liquid oxygen have a high I_{sp} as they deliver high thrust with low mass fuel consumption. In contrast, $I_{sp} = v_{ex}/g_o$ is about ten times lower for cold gas thrusters as they give much lower thrust for equal propellant weight. The underlying physical basis is the high exhaust speed v_{ex} achieved in chemically reactive liquid mixtures compared with those of cold gas expansion in a nozzle.

Increasing the specific impulse and/or the thrust delivered leads to reductions in the required propellant weight that have a direct impact in the satellite design, allowing heavier payload masses m_f and longer operational time in orbit. New technologies that increase the propellant exhaust speed have a direct impact in the economic competitiveness of satellite systems.

6.4 Electric propulsion principles

The thrust $F = \dot{m}_p\, v_{ex}$ delivered by a propulsive system is achieved by the transformation of energy (chemical, electric power, etc) into kinetic energy power [4],

$$P_b = \frac{dE_{kin}}{dt} = \frac{1}{2}\dot{m}_p\, v_{ex}^2 = \frac{F^2}{2\,\dot{m}_p} \tag{6.5}$$

of the gas, liquid or solid propellant substance.

Achieving high values of P_b in equation (6.5) in conventional chemical thrusters basically depends on the nature of the propellant substance and requires high mass flow rates $\dot{m}_p$ and/or elevated gas temperatures T. The kinetic energy power P_b comes from the transformation of thermal energy of molecules of mass m_a into a gas stream of $v_{ch} \sim \sqrt{T/m_a}$ exhaust velocity using a solid nozzle.

The in-space electric propulsion systems make use of physical processes involving electric power to increment the exhaust velocity of a plasma flow and/or to heat a gas stream. These engines operate in a high vacuum environment and the diagram of figure 6.2(b) shows schematically the basic elements of an EP system. The central block symbolizes an idealized electric engine that delivers a $F = \dot{m}_p\, v_{ex}$ thrust by transforming the electric power W into the kinetic energy power (6.5).

To this aim the electric energy is transformed into gas thermal energy raising its temperature by collisions of charged particles with neutral atoms and/or the propellant is directly heated. The thrust is delivered along a fixed direction by the hot gas exhaust expansion in a conventional solid nozzle.

Alternatively, propellant molecules can be ionized and ions of mass m_i accelerated to supersonic velocities $v_{ex} \sim \sqrt{2\, e\, \phi / m_i}$ by externally applied and/or self-consistent electric potential ϕ. The resulting ion-electron *plasma stream* that can also be focused on the direction of thrust using electrically biased electrodes and/or along the field lines of *magnetic nozzles*. No solid walls can be used in this case since ions and electrons impacting solid surfaces can recombine and/or produce additional charged particles as discussed in section 6.4.

In both cases the propellant exhaust speed is essentially independent of its chemical nature, contrary to the conventional thrusters of table 6.2. The typical exhaust velocity range of electric propulsion systems is indicated by an arrow in figure 6.2(a).

The specific performances of some electric thrusters are in table 6.3, and can be compared with those of table 6.2 for conventional chemical propulsion. The EP systems are characterized by high specific impulses $I_{sp} = v_{ex}/g_o$ and low thrusts F, whereas conventional propulsion has lower I_{sp} but delivers higher thrusts. The specific impulses in table 6.2 are below $I_{sp} = 560$ s that gives the maximum speed of $v_{ex} \sim 5.5$ km s^{-1} in figure 6.2(a). The I_{sp} values are much higher for most electric thrusters in table 6.4 reflecting the larger achievable exhaust speeds of the propellant.

A simple calculation shows the limited thrust delivered with EP systems, as high electric power and currents are required to obtain significant impulses. Under ideal

Table 6.3. Performance data and standard propellant of several types of electrical thrusters based on references [3, 5, 6]. The burning time t_b is indicated in hours (h) or is not available (n/a). The acronyms are: ECR, electron cyclotron resonance thruster; HEMPT, highly efficient multistage plasma thruster and MPD magnetoplasmadynamic thruster.

Thruster	I_{sp}(s)	Thrust (N)	Power (kW)	τ_b(s)	Propellant
Electrothermal					
Resistojet	100–300	0.2	0.1–1	~h	N_2, hydrazine, Xe
Arcjet	500	0.1	1	1000h	Hydrazine, ammonia
Helicon	~500	0.001	0.1	n/a	Ar, Kr
ECR	1000	0.001	0.03	n/a	Xe
Electrostatic					
Gridded ion engine	3000	10^{-3}–0.1	0.1–5	30 000 h	Xe
Hall thruster	1500	0.01–1	0.2–20	10 000h	Xe
HEMPT	1500–3000	0.04–0.1	5–6	9000 h	Xe
Electromagnetic					
MPD thruster	1000–10 000	0.5–50	100–10^3	1000 h	Ar, H_2, Li

Table 6.4. The characteristics of electric power sources employed in space from reference [3] where RTG is the radioisotope thermoelectric generator. The constant α_w is called *specific power* of the equipment.

	Solar panels	Batteries	RTG
Power (kW)	1–15	0.1–100	0.1–1
Efficiency η_e (%)	20–30	90	8
α_w (W kg^{-1})	150–300	100–1000	5

conditions, the maximum thrust values given by the ion beam current I_b of single charged ions accelerated by the electric potential drop ϕ can be estimated as,

$$F = \sqrt{\frac{2\,m_i}{e}}\; I_b\, \sqrt{\phi} \tag{6.6}$$

where m_i is the ion mass and e the elementary charge. The acceleration of $I_b = 1.5$ A current of xenon ions through a 1.5 kV voltage drop requires 2.25 kW of electric power and only 95.9 mN are obtained using equation (6.6). This value drops to 54.6 mN for argon gas because of its lower atomic mass. Therefore, to achieve impulses as low as 0.01 N would require space systems to cope with high voltages and large currents.

The electric energy that can be made available in the satellite and the weight of the power supply source are two criteria of relevance for EP systems. Table 6.4 shows the characteristics of electric power sources usually employed in spacecraft. The efficiency η_e for a solar panel is the fraction of sunlight energy[1] converted into electricity which is below η_e for batteries. Both are used in combination to power satellites in Earth orbit, whereas radioisotope thermoelectric sources (RTG) are less efficient and mainly used in deep space exploration.

In conventional chemical propulsion the weight of propellant decreases in time, whereas in EP systems the energy source remains constant, so its advantages can be outweighed by the large mass of the electric power equipment required. The *specific power* coefficient α_w in table 6.4 is the estimated rate between the output power delivered in watts and the average weight of the power supply in kilograms. For a typical $\alpha_w = 225$ W kg^{-1} value and 2.25 kW we need 10 kg of power equipment to deliver 95.9 mN as in the above example. As a consequence, applications of in-space electric propulsion are restricted to low thrust until new advances in electric power technology improve the basic data of table 6.4.

Chemical thrusters have burning times τ_b from minutes to one hour in table 6.2, whereas most electric propulsion systems in table 6.3 have τ_b between 10 000 hours (427 days) and 30 000 hours. This last figure represents more than three years of continuous operation and therefore the *total impulse* $I_t = F \times \tau_b$ delivered can be comparable to the thrusters of table 6.2. Therefore, electric propulsion can be

[1] It is usually estimated as 1360 W m^{-2} for Earth orbit.

advantageous for long time maneuvers, as the amount of propellant required is much lower.

For example, the 400 N hydrazine monopropellant thruster used by the Ariane 5 G upper stage consumes 300 kg of propellant in 850 seconds to deliver a maximum impulse $I_t = 1.88 \times 10^3$ N. The Hall effect thruster in table 6.3 reaches this total impulse in 56 hours using much lower propellant weight.

6.4.1 Efficiencies

The kinetic energy P_b per time unit delivered to the ion beam of figure 6.2(b) by the thruster is the *beam* or *kinetic power* introduced in equation (6.5),

$$P_b = \frac{1}{2}\, F \times v_{ex} = \frac{F^2}{2\, \dot{m}_p}$$

When plasma acceleration is the working principle of a thruster, not all particles in the propellant gas are ionized and neutral atoms do not significantly contribute to either P_b or the thrust F. We can introduce the *mass utilization efficiency* as the ratio,

$$\eta_m = \frac{\dot{m}_i}{\dot{m}_p} \leqslant 1$$

which measures the efficiency of the propellant ionization. This parameter is $\eta_m = 1$ in the limit cases of fully ionized gas or when the thrust is not produced by charged particle acceleration. As the propellant mass rate m_p is usually simpler to control than $\dot{m}_i$ we have,

$$F = \dot{m}_i\, v_{ex} = \eta_m\, \dot{m}_p\, v_{ex} \quad \text{and,} \quad P_b = \frac{F^2}{2\, \eta_m\, \dot{m}_p}$$

and the specific impulse can be also corrected since,

$$I_{sp} = \frac{F}{\dot{m}_p\, g_o} = \eta_m \frac{v_{ex}}{g_o} \tag{6.7}$$

The electrical efficiency of a thruster can be characterized by the *thrust* or *electrical efficiency* parameter [3, 7],

$$\eta_e = \frac{P_b}{W_T} < 1$$

where W_T is the total electrical input power of the thruster. This ratio measures the efficiency in the conversion of electric power in propellant kinetic energy and we have $\eta_e < 1$ since this conversion is not perfect. We can write,

$$\frac{F}{W_T} = \eta_e \times \frac{F}{P_b} = \eta_e \times \frac{2}{v_{ex}}$$

and using $v_{ex} = I_{sp} \times g_o / \eta_m$

$$\frac{F}{W_T} = \eta_e \, \eta_m \times \frac{2}{I_{sp} \, g_o} \tag{6.8}$$

We can introduce a *total efficiency* $\eta_T = \eta_m \times \eta_e$. In figure 6.3 are represented the values of equation (6.8) as a function of the specific impulse I_{sp} for different rations η_T together with the actual characteristics of different propulsive systems. Equation (6.8) can be used to relate the electric power required by the thrusters of figure 6.3. For example a 100 mN impulse from an HET thruster with $\eta_T = 0.5$ and $I_{sp} \simeq 2000$ s requires about 2 kW power.

The solid red line of $\eta_T = 100\%$ in figure 6.3 indicates the *maximum ideal efficiency* when all propellant neutral gas is fully ionized ($\eta_m = 1$) and additionally all available electric power W_T is transformed into kinetic energy ($\eta_e = 1$) of ions in the plasma jet. The solid lines below correspond to the decreasing efficiencies in equation (6.8). At the left of the dotted vertical line, which marks limit ($I_{sp} \simeq 560$ s) lies the classical chemical propulsion as well as arcjets and resistojets. Most commercial electric thrusters (GIE and HET) have specific impulses over 1000 s and efficiencies in the order of $\eta_T \sim 50\%$ approximately.

6.5 Modified Tsiolkovsky equation

Next, we will consider the corrections introduced by the weight of electric power equipment in masses ratio in the Tsiolkovsky equation. The propellant consumption m_{po} along the burning time τ_b is,

$$m_{po} = \dot{m}_p \times \tau_b = \frac{\dot{m}_i}{\eta_m} \times \tau_b = \frac{\tau_b}{\eta_m} \times \frac{F}{v_{ex}}$$

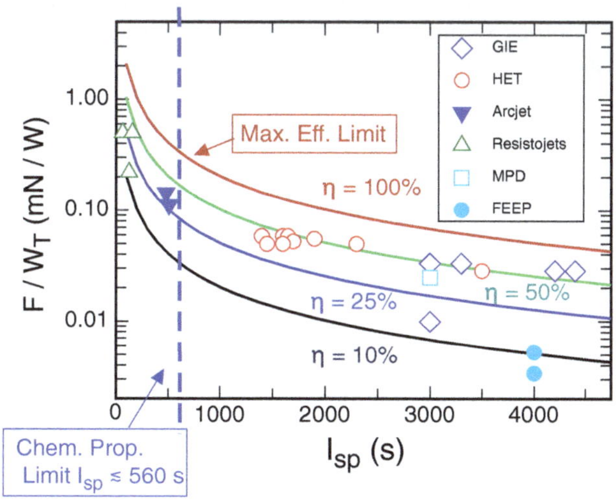

Figure 6.3. Equation (6.8) for different total efficiency rates η_T as a function of the specific impulse I_{sp}. The symbols corresponds to actual values of commercial electric thrusters of table 6.5.

replacing v_{ex} using equation (6.7),

$$m_{po} = \frac{F\,\tau_b}{g_o} \times \frac{1}{I_{sp}}$$

The amount of fuel m_{po} decreases with I_{sp}, however, we need to consider also the weight of the electric power equipment that is constant in time. This mass can be considered approximately proportional to the electrical power W supplied, $m_w = \alpha_w\, W$ where the constant α_w is called *specific power* and its typical values are in table 6.4. We can write,

$$m_w = \alpha_w\, W = \alpha \times \frac{P_b}{\eta_e} = \frac{\alpha_w}{2\,\eta_e}\, F \times v_{ex}$$

where the electric power W is transformed into kinetic energy with the $\eta_e < 1$ efficiency factor. Introducing $v_{ex} = g_o\, I_{sp}/\eta_m$ we find,

$$m_w = \frac{\alpha_w\, g_o\, F}{2\,\eta_T} \times I_{sp}$$

the weight of the power equipment increases with the specific impulse. Since the electric thruster needs both propellant and power supply to operate for the initial total mass of the spacecraft m_o we have,

$$m_o = m_f + m_{po} + m_w = m_f + \left(\frac{F\,\tau_b}{g_o}\right) \times \frac{1}{I_{sp}} + \left(\frac{\alpha_w\, g_o\, F}{2\,\eta_T}\right) \times I_{sp} \qquad (6.9)$$

where m_f is the constant final mass delivered after the maneuver.

For a constant delivered thrust F and burning time τ_b the second term in equation (6.9) decreases with $I_{sp} = \eta_m\, v_{ex}/g_o$, whereas the second addition linearly grows with the specific impulse. The amount of required propellant reduces as v_{ex} grows but higher exhaust velocities require heavier power supplies. Therefore, an optimal I_{sp} value for m_o exists, as shown in figure 6.4, and the minimum mass corresponds to,

$$\left(I_{sp}\right)_{\min} = \frac{1}{g_o}\sqrt{\frac{2\,\eta_T\, F\,\tau_b}{\alpha_w}}$$

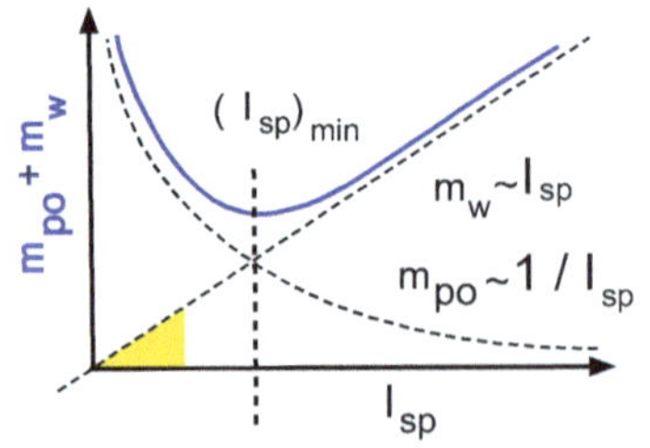

Figure 6.4. The mass $m_{po} + m_w$ of equation (6.9) as a function of I_{sp}.

Next, we study the effect of this optimal value of in the Tsiolkovsky equation (6.2) that relates the increase Δv in the spacecraft velocity and the ratio m_f/m_o between the delivered mass m_f and m_o the initial mass of the system. We have,

$$m_o = m_f + m_{po} + m_w = m_f + m_{po} + \alpha_w \, \eta_T \, P_b$$

and using $P_b = \dot{m}_p \, v_{ex}^2/2$ and $\dot{m}_p = m_p/\tau_b$ then,

$$m_o = m_f + m_{po} + \alpha_w \, \eta_T \, \frac{m_{po} \, v_{ex}^2}{2 \, \tau_b}$$

The ratio $u_b = (2 \, \tau_b/\alpha_w \, \eta_T)^{1/2}$ has the physical units of a velocity and we can write,

$$m_o - m_f = m_{po}\left[1 + \frac{v_{ex}^2}{u_b^2}\right] \quad \text{and,} \quad m_{po} = \frac{m_o - m_f}{1 + (v_{ex}/u_b)^2} \tag{6.10}$$

The mass of the electric power equipment is also,

$$m_w = \alpha_w \, \eta_T \, P_b = \alpha_w \, \eta_T \, \frac{m_{po} \, v_{ex}^2}{2 \, \tau_b}$$

and using the previous expression for m_{po} it can be cast as,

$$m_w = \alpha_w \, \eta_T \frac{v_{ex}^2}{2 \, \tau_b} \frac{m_o - m_f}{1 + (v_{ex}/u_b)^2} = \left(\frac{v_{ex}}{u_b}\right)^2 \frac{m_o - m_f}{1 + (v_{ex}/u_b)^2}$$

Finally we obtain,

$$m_w = \frac{m_o - m_f}{1 + (u_b/v_{ex})^2} \tag{6.11}$$

Equation (6.11) can be combined with equation (6.2) where the final mass is $m_f + m_w$ to incorporate the power equipment,

$$\exp\left(\frac{\Delta v}{v_{ex}}\right) = \frac{m_o}{m_f + m_w} = \frac{m_o \, (1 + u_b^2/v_{ex}^2)}{m_o + m_f \, (u_b^2/v_{ex}^2)} = \frac{(1 + v_{ex}^2/u_b^2)}{v_{ex}^2/u_b^2 + m_f/m_o}$$

and finally we can write,

$$\frac{\Delta v}{u_b} = \left(\frac{v_{ex}}{u_b}\right) \ln\left[\frac{(1 + v_{ex}^2/u_b^2)}{v_{ex}^2/u_b^2 + m_f/m_o}\right] \tag{6.12}$$

This expression relates the increments in the spacecraft speed Δv normalized to the characteristic speed $u_b = (2 \, \tau_b/\alpha_w \, \eta_T)^{1/2}$ with scaled exhaust speed v_{ex}/u_b and the ratio between the final m_f and initial m_o masses of the system. The characteristics of the electric power equipment is incorporated in u_b through the coefficient α_w in table 6.4 as well as the total efficiency η_T of the thruster.

Equation (6.12) is represented in figure 6.5 as a function of the normalized exhaust velocity v_{ex}/u_b for different $r = m_f/m_o$ mass ratios. The increments Δv are

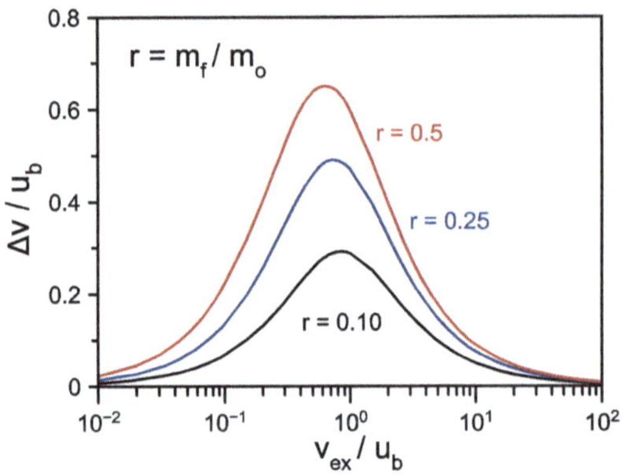

Figure 6.5. Normalized increments $\Delta v/u_b$ (equation 6.12) as a function of v_{ex}/u_b for different $m_r = m_f/m_o$ values.

not equally distributed and it is desirable to operate close to $v_{ex}/u_b \sim 1$ in the vicinity of the sharp maxima. This ratio requires a $\tau_b \sim$ 13 days burning time for a thruster with $\eta_T \sim 0.5$ (see figure 6.3) and propellant exhaust velocity of $v_{ex} \sim 20$ km s^{-1} and $\alpha_w \sim 225$ W kg^{-1} (see table 6.4).

Consequently, the efficient use of electric thrusters requires long times of operation to maximize the low thrust delivered and the need for power equipment that is an integral part of the propulsive system.

6.6 Classification of electric thrusters

The scheme of table 6.5 shows the electric thrusters can be roughly classified in accordance with the physical principle employed for the acceleration of propellant as *electrothermal*, *electrostatic* and *electromagnetic*.

When the working principle of the thruster is the acceleration of a plasma stream, the system usually needs at least one cathode to provide electrons for the ionization of the neutral gas and also for the space charge neutralization of the ion beam [8] (see section 3.4.3). Different electrons sources have been employed, such as thermionic electron emitters or dispenser cathodes, but hollow cathodes are today a standard choice for space applications [7].

The *hollow cathodes* produce the fractional ionization of neutral gas flow using thermionic electrons from a low work function electrode called *insert*, which is heated by an electric current. The plasma is produced by a DC electric discharge at its exit section and it can be seen in operation in the photographs of figures 6.7–6.11. This element is the electron emitting-cathode indicated in the schematic diagrams 6.8, 6.10 and 6.12 of plasma thrusters.

6.6.1 Electrothermal

The *electrothermal* propulsion makes use of an electric current to increase the temperature of the propellant gas by ohmic heating and/or electron-neutral collisions. The kinetic energy of molecules becomes higher and thrust is delivered

Table 6.5. Non-exhaustive classification of EP thrusters where m_a and m_i are, respectively, the masses of molecule and ion. The electric potential is ϕ, the neutral gas or ion temperature is T and J the electric current density. The acronyms are: ECR, electron cyclotron resonance thruster; GIE, gridded ion engine; FEEP, field emission electric propulsion; HET, Hall effect thruster, HEMPT, highly efficient multistage plasma thruster. The applied-field (AF) and self-field (SF) are variants of the magnetoplasmadynamic MPD thruster.

Physical principle			Thruster
Electrothermal	$v_{ex} \sim \sqrt{\frac{T}{m_a}}$	Solid nozzle	Arcjet Resistojet
		Magnetic nozzle	Helicon ECR
Electrostatic	$v_{ex} \sim \sqrt{\frac{2e\phi}{m_i}}$	Externally applied electric field	GIE Colloid FEEP
		Self-consistent electric field	HET HEMPT Multi-Cusp
Electromagnetic	$v_{ex} \sim \frac{J}{\dot{m}}$	Steady plasma flow	AF-MPD SF-MPD
		Unsteady plasma flow	Pulsed plasma

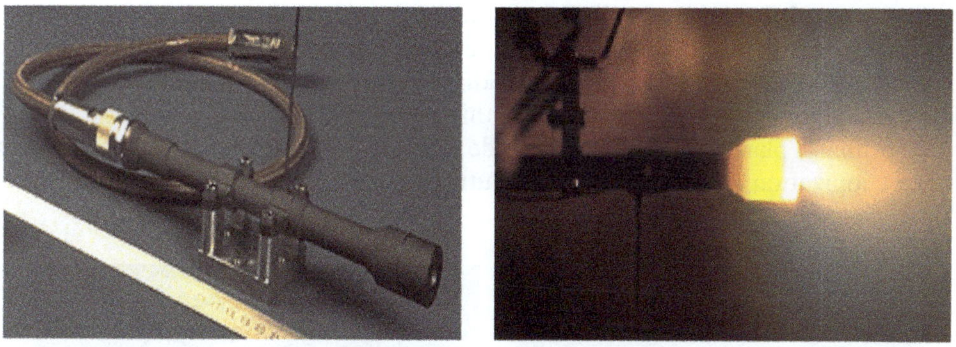

Figure 6.6. The ARTUS-IM3 arcjet on the left and operating into a vacuum chamber on the right picture. (Courtesy Institute of Space Systems, Sttutgart University, Germany.)

to the spacecraft by the expansion through a solid nozzle of this hot neutral gas, usually with a low ionization degree.

The *resistojet* makes use of a resistor to heat the gas and this energy transfer mechanism has a low efficiency, leading to reduced specific impulses of $I_{sp} \sim 100$–300 s in table 6.3, below the limit of figure 6.3 for chemical propulsion.

In the *arcjet* thruster shown in figure 6.6 the heating electric current flows through the propellant and collisions of electrons with neutral gas atoms increase the gas temperature. The energy transfer requires elevated mass flow rates to increase the electron-neutral ν_{ea} collision frequencies which give $I_{sp} \sim 500$ s, close to the limit of figure 6.3.

Microwave and radio-frequency plasma thrusters (Helicon and electron cyclotron resonance, ECR in table 6.5) are currently included in this categories and are still the subject of study [3]. Different physical mechanisms (helicon wave, electron cyclotron resonance, etc) transfer the energy of an electromagnetic wave to the plasma electrons to increment their kinetic energy and therefore the ionization rate of the propellant gas. Since the motion of charged particles can be restricted along the magnetic field lines, the imparted momentum increases when plasma jet expands through a *magnetic nozzle* configuration [9].

These microwave and RF thrusters essentially rely on the ambipolar electric field introduced in section 3.2.3 to drive the ions that deliver thrust. The advantage is the absence of plasma exposed electrodes and in the case of ECR no cathode is needed to ionize the neutral gas. Therefore, the erosion and wearing of thruster materials is much lower compared with resistojets and arcjets.

6.6.2 Electrostatic

The *electrostatic* thrusters make use of electric power to ionize the propellant neutral gas to produce an electron–ion plasma and momentum is directly imparted to heavy ions by a stationary electric field. As shown in equation (6.6), ions can reach high velocities moving in the field produced by electrodes (usually metallic grids) connected to high voltage power supplies.

In the *gridded ion engines* (GIE, also called Kaufman thrusters) of figure 6.7 the electrons from a cathode placed inside the discharge produce the impact ionization of the propellant gas atoms. Its longitudinal cross sectional diagram of figure 6.8 illustrates the steady state operation of this plasma thruster.

A primary plasma is produced inside the discharge chamber by the electron impact ionization of the propellant neutral gas by free electrons from a hollow

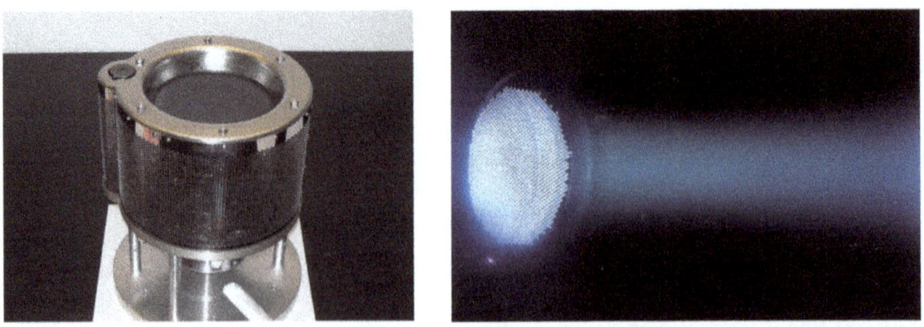

Figure 6.7. The 10 cm diameter T5 ion engine employed in the GOCE mission (left) and this thruster operating in a vacuum tank on the right side. (Courtesy QinetiQ Ltd.) The hollow cathode exit section is the small circle on its surface on the left and in the photograph of the thruster operation on the right, is the small bright blue dot below.

cathode. These are also confined by the magnetic field lines configuration inside the ionization chamber to increase the ionization rate of the neutral gas.

Two grids first extract and then accelerate these ions to high speeds modulated by the electric potential of the acceleration grid of figure 6.8. An additional electron-emitting cathode disposed past the grids provides electrons to neutralize the ion outflow, as was discussed in section 3.4.2. The third external (decelerating) grid is biased at a negative potential to prevent the electrons from the neutralizer cathode moving towards the thruster, since this is a failure mechanism that deteriorates the grids.

The primary plasma production, ion acceleration and space charge neutralization are independent processes in GIE thrusters. Their specific impulses are in the order of $I_{sp} \sim 3000$ s in table 6.3 and deliver $F = 0.001$–1.0 N impulses.

These are regulated by the electric voltage applied to the acceleration grid. The GIE is a mature technology that has been successfully employed in space [1] in the GOCE mission in Earth orbit or the deep space Beppi-Colombo mission to Mercury.

Other thrusters mentioned in table 6.3 make use of similar schemes where electrodes are connected to high voltages. Highly charged micro-droplets of polar liquids accelerated are under kV-range voltage drops are employed by colloid thrusters to impart thrust.

Similarly, the *field emission electric propulsion* (FEEP) produces the electric field ionization of a liquid metal (cesium, indium or mercury) and ions are later accelerated by an externally imposed strong electric field. The advantage is their small size and low power consumption that make them suitable for micro-satellite propulsion and precision motions since they deliver thrusts in the 1–100 μN range with high specific impulses, as indicated in figure 6.3.

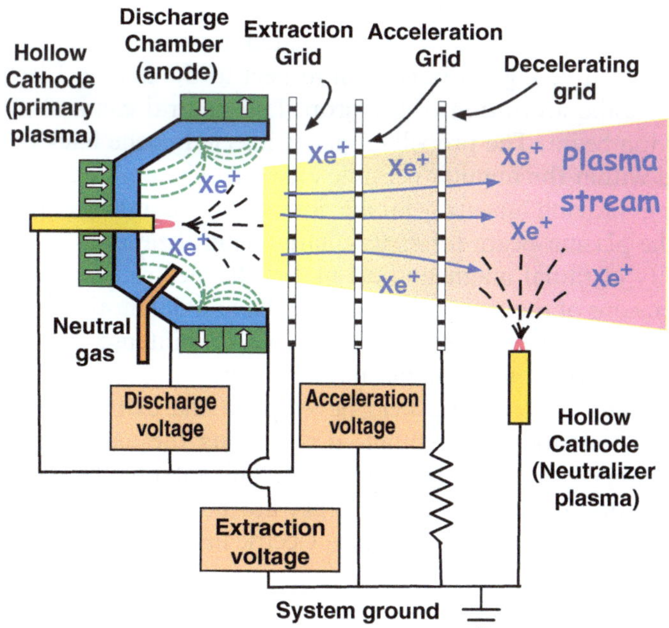

Figure 6.8. Schematic diagram of a gridded ion engine (GIE) as in figure 6.7. The polarity of permanent magnets in green is indicated by the arrows.

Figure 6.9. The Hall effect thruster PPS 1340-E on the left and operating in a vacuum tank in the right photograph. (Courtesy Safran-Snecma.) The hollow cathode is the cylinder attached to the equipment at its upper left and in operation is the bright dot of the right photograph.

The non-equilibrium discharge plasmas can develop internal self-consistent electric fields that can accelerate ions to velocities well over the limit for chemical propulsion indicated in figure 6.2(b). The *Hall effect thruster* (HET) (also called *closed-drift* thruster) of figure 6.9 and the *highly efficient multistage plasma thruster* (HEMPT) shown in figure 6.11 make use of this principle.

The cross sectional scheme of the HET thruster along its longitudinal direction is shown in figure 6.10. The gridless HET has a cylindrical ring shaped ionization chamber usually made of non conductive materials[2] with an annular anode place at its closed end. A radial magnetic field inside is generated by a first pole configuration about the central axis of the system. The second counter-pole magnet surrounds the ionization chamber on the outside.

The only cathode in the Hall thruster is located outside the plasma chamber and is connected to a positive high voltage with respect to the anode. Electrons from this cathode produce the ionization of the propellant neutral gas and also the neutralization of the ion beam. The propellant gas is injected at the back end of the anode and is ionized within the annular ceramic channel. The emitted electrons spread in two directions, moving inwards to the ionization chamber and also along the direction of the plasma beam for space-charge neutralization.

The intensity of radial magnetic field makes the electron Larmor radius smaller than the dimensions of the ionization channel. Their axial motion is inhibited as they move towards the anode since they must cross this intense radial magnetic field formed at the open section of the plasma chamber.

A self-consistent axial electric field develops at the exit section of the ceramic channel since the cathode bias voltage drops along the axial direction where the electrons are trapped by this *magnetic barrier*. This intense electric field accelerates the ions up to speeds of 20–40 km s^{-1} that are non-magnetized because of their heavier mass. This combination of the axial electric field $\boldsymbol{E}$ and radial magnetic field $\boldsymbol{B}$

[2] The discharge chamber is made of metallic materials in the variant of HET called *thruster with anode layer* (TAL), which has a shorter ionization chamber.

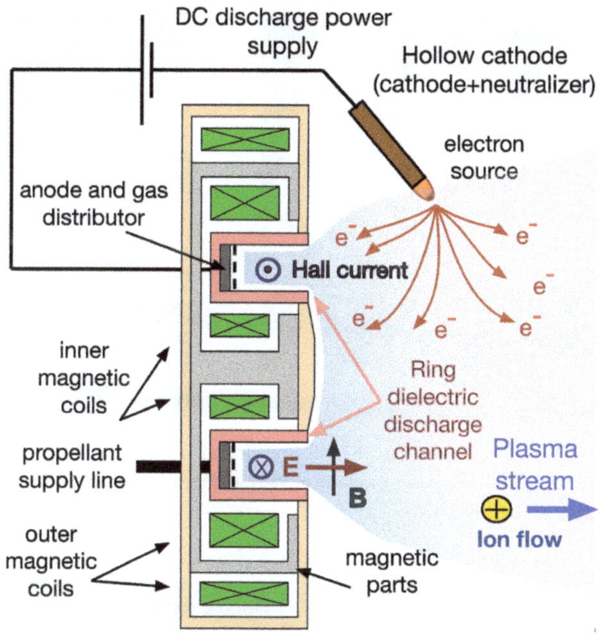

Figure 6.10. Schematic diagram of a Hall effect thruster (HET) as in figure 6.9.

produces an annular Hall current in the $\boldsymbol{E} \times \boldsymbol{B}$ direction that can be seen in figure 6.9, confines the plasma and enhances the ionization rate, as the electrons drift circularly within the annular plasma chamber. Figures 6.7 and 6.9 show the lower divergence HET plasma streams compared with GIE where the ion motion is better focused.

This basic scheme is similar for all different HET configurations and a number of satellites have been equipped with them for in-orbit operation and station keeping [1]. They have a wide range of $F \sim 0.01$–1 N with specific impulses of $I_{sp} \sim 1500$ s.

The HEMPT thruster can be considered one of the multi-cusped field thruster category also mentioned in table 6.3, and its cross sectional diagram is shown in figure 6.11. It has a cylindrical discharge chamber made of a non-conductive material with the electron source placed in front of its open side, as in the HET designs. The anode is placed at the closed back end of the ionization chamber on the external surface of which ring shaped permanent magnets with alternate polarities are mounted. One or more annular electrodes are successively distributed longitudinally over the surface of the plasma chamber.

The plasma is produced by biasing the back anode with respect to the external cathode, which provides electrons for neutral gas ionization and neutralization of the ion beam. Electron confinement by the longitudinal magnetic field along the axis of symmetry increases the electron impact ionization rate in the chamber. The ions are accelerated by the multistage electric field established by biasing the electrodes with electric potentials of increasing values between the anode and the cathode potentials.

The divergence of the HEMPT plasma stream is also higher than in GIE as shown in photograph 6.11, and its specific impulses are $I_{sp} \sim 1500$–3000 s with

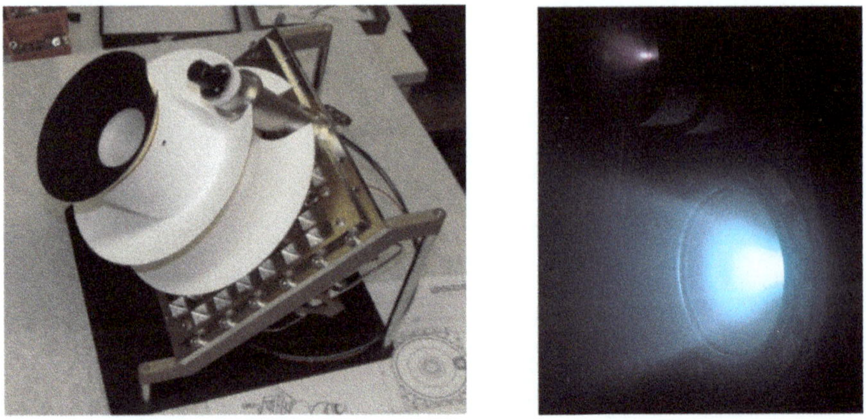

Figure 6.11. The HEMPT plasma thruster at the left and operating into a vacuum tank in the right picture. (Courtesy of Thales Group.) The hollow cathode is placed at the upper right of the equipment and it is the bright dot in photograph of the thruster operation.

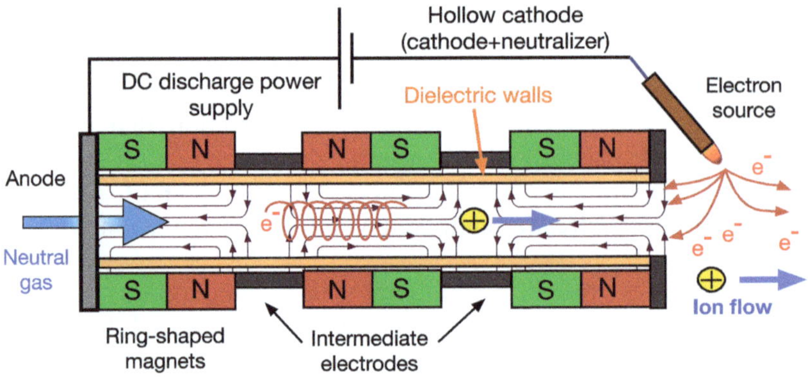

Figure 6.12. Schematic diagram of the HEMPT plasma thruster of figure 6.11.

$F \sim 0.04$–0.1 N impulses. This thruster has reached an important level of development but it has not been tested in orbit so far.

6.6.3 Electromagnetic

The propellant gas is also ionized in the *electromagnetic* propulsion and this plasma is accelerated by the interaction of current flowing through with steady or pulsed electric and magnetic fields established by the currents themselves and/or external means.

These promising and complex devices that are still a subject of research deliver high thrusts ($F \sim 0.5$–50 N) with specific impulses in the range of 10^3–10^4 s. Their drawback is their large electric power consumption, in the range of megawatts, and their short lifetime by the degradation and evaporation of materials.

In the self-field magnetoplasmadynamic thruster (SF-MPD in table 6.5) a high discharge current $\boldsymbol{j}$ is established between two concentric electrodes. The cathode is disposed along the center line and the propellant flows parallel, whereas $\boldsymbol{j}$ is along in the radial direction. The magnetic Lorentz force $\boldsymbol{F} = \boldsymbol{j} \times \boldsymbol{B}$ resulting from the self-produced magnetic field $\boldsymbol{B}$ in the azimuthal direction accelerates the plasma stream to high velocities.

The applied-field (AF-MPD) thrusters have coils surrounding the thruster to produce an additional magnetic field parallel to the plasma chamber. This applied magnetic field affects the physical processes in the thruster discharge chamber and also the diverging field lines in the plasma expansion region have a geometry similar to a *magnetic nozzle*. Then, the plasma becomes confined along the radial direction and its expansion contributes to the thrust.

6.7 Commentaries and further reading

This chapter is an introduction and not a systematic account of all propulsive systems that use electrical energy. The field is rapidly evolving and new ideas appear constantly. For more information, the interested reader is referred the comprehensive reviews [3, 10] and [11], that offer an excellent up-to-date view of the current state of this technology, as well as [12] and [13]. The book [7] contains an excellent introduction to electric propulsion, specifically to ion engines and Hall thrusters and is more up to date than the classic book [14]. An overview of thermal arcjet thrusters is in the review paper [15].

The electron sources are crucial for electric propulsion systems, the hollow cathodes are today a standard for space applications and are discussed in chapter 6 of reference [7]. Chapter 5 of the book [8] discusses the ion beam neutralization issues. Magnetic nozzles are mainly a subject of research so far and their physical models are complex; reference [9] is a comprehensive review in the context of ECR and Helicon plasma thrusters.

References

[1] List of spacecrafts with electric propulsion. Wikipedia https://en.wikipedia.org/wiki/List_of_spacecraft_with_electric_propulsion accessed on May 22, 2020

[2] Dudeck M, Doveil F, Arcis N and Zurbach S 2012 Plasma propulsion for geostationary satellites for telecommunication and interplanetary missions *IOP Conf. Ser.: Mater. Sci. Eng.* **29** 012010

[3] Mazouffre S 2016 Electric propulsion for satellites and spacecraft: established technologies and novel approaches *Plasma Sour. Sci. Technol.* **25** 1–27

[4] Sutton G P and Biblarz O 2001 *Rocket Propulsion Elements* (New York: Wiley)

[5] Boch E 2017 The HEMPT NG *Proc. of the Electric Propulsion, Innovation & Competitiveness Workshop (Madrid, Spain)*

[6] Koch N *et al* 2011 The HEMPT concept. A survey on theoretical considerations and experimental evidences *Proc. of the 32th Int. Electric Propulsion Conf. (Weisbaden, Germany)* IEPC paper 2011-236

[7] Goebel D M and Katz I 2008 *Fundamentals of Electric Propulsion. Ion and Hall Thrusters* (Hoboken, NJ: Wiley)

[8] Humphries S 2013 *Charged Particle Beams* 2nd edn (New York: Dover)

[9] Takahasi K 2019 Helicon-type radiofrequency plasma thrusters and magnetic plasma nozzles *Rev. Mod. Plasma Phys.* **3** 1–61

[10] Levchenko I, Xu S, Mazouffre S, Lev D, Pedrini D, Goebel D, Garrigues L, Taccogna F and Bazaka K 2020 Perspectives, frontiers, and new horizons for plasma-based space electric propulsion *Phys. Plasmas* **27** 020601

[11] Jahn R G and Choueiri E Y 2002 Electric propulsion *Encyclopedia of Physical Science and Technology* ed R A Mayers vol 5 3rd edn (Cambridge MA: Academic)

[12] Charles C 2009 Plasmas for spacecraft propulsion *J. Phys. D: Appl. Phys.* **42** 163001

[13] Ahedo E 2011 Plasmas for space propulsion *Plasma Phys. Control. Fusion* **53** 124037

[14] Jahn R G 2006 *Physics of Electric Propulsion* 2nd edn (New York: Dover)

[15] Wollenhaupt B, Le Q H and Herdrich G 2018 Overview of thermal arcjet thruster development *Aircr. Eng. Aerosp. Tech.* **90** 353–68

IOP Publishing

An Introduction to Plasma Physics and its Space Applications, Volume 2
Basic equations and applications

Luis Conde

Appendix A

The substantial derivative

The motion of a point charge q of mass m moving in an electromagnetic field is usually described in terms of its *Lagrangian coordinates* $(\boldsymbol{R}(t),\, t)$ where the particle initially located in t_o at the point $\boldsymbol{r}_o$ of the space moves for $t \geqslant t_o$,

$$\boldsymbol{R}(t) = \boldsymbol{r}_o + \boldsymbol{s}(\boldsymbol{r}_o,\, t) \quad \text{with velocity,} \quad \boldsymbol{V}(t) = \frac{d\boldsymbol{s}}{dt}$$

The Lorentz force (equation (2.5) of volume 1) gives the acceleration,

$$\boldsymbol{A}(t) = \frac{d\boldsymbol{V}}{dt} = \frac{q}{m}[\boldsymbol{E}(\boldsymbol{R},\, t) + \boldsymbol{V} \times \boldsymbol{B}(\boldsymbol{R},\, t)] \tag{A.1}$$

where the values of electromagnetic fields are calculated at the points $\boldsymbol{R}(t)$ along the path of the electric charge. However, the electromagnetic fields in equation (A.1) are usually expressed in *Eulerian coordinates* $(\boldsymbol{r},\, t)$ because they are solutions of the system of partial differential equations (2.6) in volume 1, and $\boldsymbol{E}(\boldsymbol{r},\, t)$ and $\boldsymbol{B}(\boldsymbol{r},\, t)$ are vectors determined at each fixed point $\boldsymbol{r}$ of space.

Sometimes the calculations become simpler when we express the charge velocity $\boldsymbol{v}(\boldsymbol{r},\, t)$ and acceleration $\boldsymbol{a}(\boldsymbol{r},\, t)$ fields similarly to $\boldsymbol{E}$ and $\boldsymbol{B}$, instead of using $\boldsymbol{R}(t)$ and $\boldsymbol{V}(t)$ as in equation (A.1). The *Eulerian* coordinate $\boldsymbol{r} = \boldsymbol{R}(t)$ of the particle is simply,

$$\boldsymbol{r} = \boldsymbol{r}_o + \boldsymbol{s}(\boldsymbol{r}_o,\, t) \quad \text{and at this point,} \quad \boldsymbol{v}(\boldsymbol{r},\, t) = \frac{d\boldsymbol{s}}{dt}$$

the Eulerian velocity field $\boldsymbol{v}(\boldsymbol{r},\, t)$ and the Lagrangian velocity $\boldsymbol{V}(t)$ are equal. For the increment $\delta\boldsymbol{s}$ we have,

$$\frac{d}{dt}\boldsymbol{s}(\boldsymbol{r}_o,\, t + \delta t) = \boldsymbol{v}(\boldsymbol{r} + \delta\boldsymbol{s},\, t + \delta t) \simeq \boldsymbol{v}(\boldsymbol{r},\, t) + \frac{\partial \boldsymbol{v}}{\partial t}\,\delta t + \sum_i \frac{\partial \boldsymbol{v}}{\partial x_i}\,\delta x_i + O(\delta t^2)$$

where $\delta x_i = v_i\, \delta t$ and $i = 1, 2, 3$ are the (x, y, z) coordinates. The Lagrangian acceleration $\boldsymbol{A}(t)$ is then,

doi:10.1088/978-0-7503-3543-0ch7

$$\lim_{\delta t \to 0} \frac{1}{\delta t}\left[\frac{d}{dt}\boldsymbol{s}(\boldsymbol{r}_o,\, t + \delta t) - \frac{d}{dt}\boldsymbol{s}(\boldsymbol{r}_o,\, t)\right] = \lim_{\delta t \to 0} \frac{1}{\delta t}[\boldsymbol{v}(\boldsymbol{r} + \delta \boldsymbol{s},\, t + \delta t) - \boldsymbol{v}(\boldsymbol{r},\, t)]$$
$$= \frac{\partial \boldsymbol{v}}{\partial t} + \sum_i v_i \frac{\partial \boldsymbol{v}}{\partial x_i}$$

equal to the Eulerian acceleration $\boldsymbol{a}(\boldsymbol{r},\, t)$ given by the velocity field $\boldsymbol{v}(\boldsymbol{r},\, t)$. Using the nabla operator $\nabla_{\mathrm{r}} = \boldsymbol{i}\, \partial/\partial x + \boldsymbol{j}\, \partial/\partial y + \boldsymbol{k}\, \partial/\partial z$ the last term can be cast as,

$$\boldsymbol{a}(\boldsymbol{r},\, t) = \frac{\partial \boldsymbol{v}}{\partial t} + \sum_i (\boldsymbol{v} \cdot \nabla_{\mathrm{r}})\, \boldsymbol{v} \tag{A.2}$$

which connects the Lagrangian and Eulerian descriptions as discussed in the example of box A.1. The vector $\boldsymbol{v}$ is also equal to the Lagrangian velocity $\boldsymbol{V}(t)$ and the scalar operator,

$$\frac{D}{Dt} = \frac{\partial}{\partial t} + \sum_i (\boldsymbol{v} \cdot \nabla_{\mathrm{r}}) \tag{A.3}$$

is currently denominated *substantial, material* or *convected derivative* and applies to either scalar $T(\boldsymbol{r},\, t)$ and vector $\boldsymbol{F}(\boldsymbol{r},\, t) = F_x(\boldsymbol{r},\, t)\, \boldsymbol{i} + F_y(\boldsymbol{r},\, t)\boldsymbol{j} + F_z(\boldsymbol{r},\, t)\, \boldsymbol{k}$ functions.

Box A.1. Lagrangian and Eulerian accelerations

Figure A1 shows one particle moving along a circumference with constant radius R where $\theta(t)$ changes in time with position,

$$\boldsymbol{R}(t) = R(\cos\theta\, \boldsymbol{i} + \sin\theta\, \boldsymbol{j}) = x\, \boldsymbol{i} + y\, \boldsymbol{j}$$
$$\boldsymbol{V}(t) = R\, \dot{\theta}\, (-\sin\theta\, \boldsymbol{i} + \cos\theta\, \boldsymbol{j}) = \dot{\theta}(-y\, \boldsymbol{i} + x\, \boldsymbol{j})$$

The Lagrangian vectors $(\boldsymbol{R}(t),\, \boldsymbol{V}(t))$ on the left are equivalent to the Eulerian vector fields $\boldsymbol{r}(x,\, y)$ and $\boldsymbol{v}(x,\, y)$ in the right term. As displayed in figure A1 the positions and velocities of P are parallel to the radial $\boldsymbol{u}_r = \cos\theta\, \boldsymbol{i} + \sin\theta\, \boldsymbol{j}$ and tangential $\boldsymbol{u}_\theta = -\sin\theta\, \boldsymbol{i} + \cos\theta\, \boldsymbol{j}$ unit vectors, respectively.

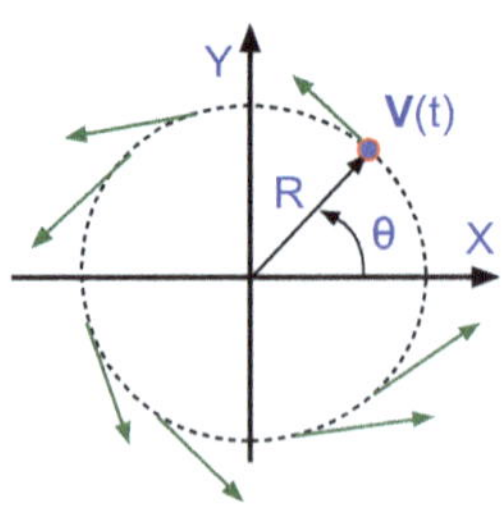

Figure A1. A point particle moving on a circle.

We can make use of equation (A.3) to calculate the acceleration,

$$\boldsymbol{a}(x, y) = \frac{\partial v}{\partial t} + v_x \frac{\partial v}{\partial x} + v_x \frac{\partial v}{\partial x} \quad \text{where,} \quad \frac{\partial v}{\partial t} = \ddot{\theta}\,(-y\,\boldsymbol{i} + x\,\boldsymbol{j})$$

$$v_x \frac{\partial v}{\partial x} = (-\dot{\theta}\, y) \times (\dot{\theta}\,\boldsymbol{j}) \ \text{ and, } \ v_y \frac{\partial v}{\partial y} = (\dot{\theta}\, x) \times (-\dot{\theta}\,\boldsymbol{i})$$

The acceleration field,

$$\boldsymbol{a}(x, y) = \ddot{\theta}\,(-y\,\boldsymbol{i} + x\,\boldsymbol{j}) - \dot{\theta}^2\,(x\boldsymbol{i} + y\,\boldsymbol{j}) = R\,\ddot{\theta}\,\boldsymbol{u}_r - R\,\dot{\theta}^2\,\boldsymbol{u}_\theta$$

is the sum of the tangential $R\,\ddot{\theta}$ and normal $R\,\dot{\theta}^2$ components to the circle of figure A1 of the acceleration of P, and both $\boldsymbol{A}(t) = d\boldsymbol{V}/dt$ and $\boldsymbol{a}(\boldsymbol{r}, t) = dv/dt$ lead to the same result.

The notation $(v \cdot \nabla_{\mathrm{r}})\,\boldsymbol{F}$ means that $(v \cdot \nabla_{\mathrm{r}})$ is applied to each of the three components F_x, F_y and F_z. For scalar functions it is simply the dot product $(v \cdot \nabla_{\mathrm{r}})\,T = v \cdot \nabla_{\mathrm{r}}\,T$ between the gradient $\nabla_{\mathrm{r}} T$ and the velocity v.

The material derivative of a physical magnitude $\boldsymbol{F}(\boldsymbol{r}, t)$ in equation (A.3) has two components, the first term $(\partial v/\partial t)$ in equation (A.2) represents the change at a fixed point in space. The second (in equation (A.2) $(v \cdot \nabla_{\mathrm{r}})\, v$) characterizes the change as the observer moves into a region where $\boldsymbol{F}(\boldsymbol{r}, t)$ is different. The reader is referred to textbooks on fluid dynamics as in reference [1] for the physical interpretation of equations (A.2) and (A.3).

Equation (A.3) can be generalized for the (scalar) velocity distribution function $f(\boldsymbol{r}, v, t)$ introduced in chapter 3 of volume 1 and chapter 2 of this volume defined in a seven-dimension space $(\boldsymbol{r}, v, t)$ where $\boldsymbol{r}$ and v and time are independent variables. The time derivative can be also computed using equation (A.3) and introducing the nabla operator $\nabla_{\mathrm{v}} = \boldsymbol{i}\ \partial/\partial v_x + \boldsymbol{j}\ \partial/\partial v_y + \boldsymbol{k}\ \partial/\partial v_z$ and,

$$\frac{Df}{Dt} = \frac{\partial f}{\partial t} + v \cdot \nabla_{\mathrm{r}} f + \boldsymbol{a} \cdot \nabla_{\mathrm{v}} f \tag{A.4}$$

This expression gives the rate of change of the scalar function $f(\boldsymbol{r}, v, t)$ in the phase space $(\boldsymbol{r}, v, t)$ where vectors v and $\boldsymbol{a}$ are the Lagrangian velocity $\boldsymbol{V}(t)$ and acceleration $\boldsymbol{A}(t)$.

Reference

[1] Batchelor G 1984 *An Introduction to Fluid Dynamics* (Cambridge: Cambridge University Press)

IOP Publishing

An Introduction to Plasma Physics and its Space Applications, Volume 2

Basic equations and applications

Luis Conde

Appendix B

The plasma kinetic equation

The kinetic model considers the effects of motion of particles in the physical description of a neutral gas or plasma. As we have seen in chapter 3 of volume 1 and chapter 2 in this volume, the macroscopic physical properties are calculated from statistical averages over a huge number of particles, which are described as points with mass and electric charge neglecting quantum mechanical effects.

In this appendix we indicate the outline of the derivation of the fundamental plasma kinetic equation (usually also called the Boltzmann equation) starting from the exact description of individual motions of plasma particles. Such a microscopic approach is obviously impractical but is on the basis of statistical models for physical systems composed of a huge number of interacting particles. We will mainly concentrate on charged particles but the basic results also apply to neutral atoms and molecules.

B.1 The Klimontovich equation

The Klimontovich equation provides us with an exhaustive description of a system of interacting particles. Each plasma particle, ions, electrons and neutral atoms ($\alpha = i, e, a$) with electric charge $q_\alpha = \pm e, 0$ moves with velocity $\boldsymbol{V}_\alpha(t)$ along the deterministic trajectory

$$\boldsymbol{R}_\alpha(t) = X_\alpha(t)\,\boldsymbol{i} + Y_\alpha(t)\,\boldsymbol{j} + Z_\alpha(t)\,\boldsymbol{k}$$

in a vacuum. We consider a six-dimensional *phase space* $(\boldsymbol{r}, \boldsymbol{v})$ or (x, y, z, v_x, v_y, v_z) where position and velocity components are the coordinates $(\boldsymbol{R}_\alpha(t), \boldsymbol{V}_\alpha(t))$ of one particle. The expression,

$$\delta[\boldsymbol{r} - \boldsymbol{R}_\alpha(t)] = \delta[x - X_\alpha(t)]\,\delta[y - Y_\alpha(t)]\,\delta[z - Z_\alpha(t)]$$

where $\delta(\boldsymbol{r} - \boldsymbol{R}_\alpha)$ is the Dirac delta function[1] takes an infinite value along the particle trajectory $\boldsymbol{R}_\alpha(t)$ and is null over all other phase space $(\boldsymbol{r}, \boldsymbol{v})$ points. We also have,

$$\int_{-\infty}^{+\infty} \delta[\boldsymbol{r} - \boldsymbol{R}_\alpha(t)]\, d\boldsymbol{r} = \int_{-\infty}^{+\infty} \delta[x - X_\alpha(t)]\, \delta[y - Y_\alpha(t)]\, \delta[z - Z_\alpha(t)]\, dx\, dy\, dz = 1$$

as well as the similar integral for $\delta[\boldsymbol{v} - \boldsymbol{V}_\alpha(t)]$. Therefore, the normalized probability density[2] *for one point particle* $f_{1p}^{m}(\boldsymbol{r}, \boldsymbol{v}, t)$ in the phase space $(\boldsymbol{r}, \boldsymbol{v})$ can be expressed as,

$$f_{1p}^{m}(\boldsymbol{r}, \boldsymbol{v}, t) = \delta[\boldsymbol{r} - \boldsymbol{R}_\alpha(t)]\, \delta[\boldsymbol{v} - \boldsymbol{V}_\alpha(t)]$$

and $f_{1p}^{m}(\boldsymbol{r}, \boldsymbol{v}, t)\, d\boldsymbol{r}\, d\boldsymbol{v}$ is null except along the positions $\boldsymbol{R}_\alpha(t)$ and velocities $\boldsymbol{V}_\alpha(t) = d\boldsymbol{R}_\alpha/dt$ of the particle. We can generalize this expression for a system of point particles labeled as $i = 1, 2, \ldots, N_\alpha$ with positions $\boldsymbol{R}_i(t)$ and velocities $\boldsymbol{V}_i(t)$ that are distributed within the six-dimensional $(\boldsymbol{r}, \boldsymbol{v})$ phase space as,

$$f_{\alpha}^{m}(\boldsymbol{r}, \boldsymbol{v}, t) = \sum_{i=1}^{N_\alpha} \delta[\boldsymbol{r} - \boldsymbol{R}_i(t)]\, \delta[\boldsymbol{v} - \boldsymbol{V}_i(t)] \tag{B.1}$$

This *microscopic distribution function* (indicated by the superscript m) is the sum of Dirac delta functions following the detailed trajectory $(\boldsymbol{R}_i(t), \boldsymbol{V}_i(t))$ of each α particle. The integral in the phase space of equation (B.1) is also normalized,

$$\int_{\infty}^{+\infty} f_{\alpha}^{m}(\boldsymbol{r}, \boldsymbol{v}, t)\, d\boldsymbol{v}\, d\boldsymbol{r} = \sum_{i=1}^{N_\alpha} \int_{-\infty}^{+\infty} \delta[\boldsymbol{r} - \boldsymbol{R}_i(t)]\, \delta[\boldsymbol{v} - \boldsymbol{V}_i(t)]\, d\boldsymbol{v}\, d\boldsymbol{r} = N_\alpha \tag{B.2}$$

Thus, $dP_\alpha = f_{\alpha}^{m}(\boldsymbol{r}, \boldsymbol{v}, t)\, d\boldsymbol{v}\, d\boldsymbol{r}/N_\alpha$ represents the probability dP_α of finding one α-particle within the phase space volume $d\boldsymbol{v}\, d\boldsymbol{r}$ centered at point $(\boldsymbol{r}, \boldsymbol{v})$. This concept was previously introduced in section 3.3 of volume 1 for the Maxwell–Boltzmann distribution of atoms in a neutral gas in thermodynamic equilibrium and generalized later in chapter 2. The number density (number of particles per volume unit) in the configuration space (x, y, z) is the sum of N_α delta functions,

$$n_\alpha(\boldsymbol{r}, t) = \int_{-\infty}^{+\infty} f_{\alpha}^{m}(\boldsymbol{r}, \boldsymbol{v}, t)\, d\boldsymbol{v} = \sum_{i=1}^{N_\alpha} \delta[\boldsymbol{r} - \boldsymbol{R}_i(t)] \text{ and, } \int_{-\infty}^{+\infty} n_\alpha(\boldsymbol{r}, t)\, d\boldsymbol{r} = N_\alpha$$

is the total number of α particles.

The Lorentz force acting over an electric charge α along the trajectories $\boldsymbol{R}_i(t)$ gives the acceleration of the $i = 1, 2, \ldots, N_\alpha$ particles as,

$$\frac{d\boldsymbol{V}_i}{dt} = \frac{q_\alpha}{m_\alpha}\left[\boldsymbol{E}^m(\boldsymbol{R}_i, t) + \boldsymbol{V}_i \times \boldsymbol{B}^m(\boldsymbol{R}_i, t)\right] \tag{B.3}$$

[1] The delta function and its properties are discussed in detail in appendix II, volume 2 of the book [1].

[2] This concept can be understood as similar to the charge density $\rho_c(\boldsymbol{r}) = q\, \delta[\boldsymbol{r} - \boldsymbol{r}_0]$ of a point charge q currently employed in classical electromagnetism, but with the probability ($P = 1$) in place of q and the delta function in the six-dimensional phase space.

We recall the superscript m indicates the *microscopic* electromagnetic fields that satisfy the Maxwell equations (2.6) in volume 1 in vacuum,

$$\nabla \cdot \boldsymbol{E}^m = \frac{\rho_c^m}{\varepsilon_o} \quad \nabla \times \boldsymbol{E}^m = -\frac{\partial \boldsymbol{B}^m}{\partial t}$$
$$\nabla \cdot \boldsymbol{B}^m = 0 \quad \nabla \times \boldsymbol{B}^m = \mu_o \left(\boldsymbol{J}_c^m + \varepsilon_o \frac{\partial \boldsymbol{E}^m}{\partial t} \right) \tag{B.4}$$

These equations govern the *microscopic* electromagnetic fields $\boldsymbol{E}^m(\boldsymbol{r}, t)$ and $\boldsymbol{B}^m(\boldsymbol{r}, t)$ that are self-consistently calculated with the electric charge $\rho_c^m(\boldsymbol{r}, t)$ and current $\boldsymbol{J}_c^m(\boldsymbol{r}, t)$ densities,

$$\rho_c^m(\boldsymbol{r}, t) = \sum_\alpha q_\alpha n_\alpha = \sum_\alpha q_\alpha \int_{-\infty}^{+\infty} f_\alpha^m(\boldsymbol{r}, \boldsymbol{v}, t)\, d\boldsymbol{v} = \sum_\alpha \sum_{i=1}^{N_\alpha} q_\alpha\, \delta[\boldsymbol{r} - \boldsymbol{R}_i(t)] \tag{B.5a}$$

$$j_c^m = \sum\nolimits_\alpha q_\alpha \int_{-\infty}^{+\infty} \boldsymbol{v} f_\alpha^m(\boldsymbol{r}, \boldsymbol{v}, t)\, d\boldsymbol{v} = \sum_\alpha \sum_{i=1}^{N_\alpha} \boldsymbol{V}_i\, \delta[\boldsymbol{v} - \boldsymbol{V}_i(t)] \tag{B.5b}$$

Here we made use of the property of the delta function $x\, \delta[x - y] = y\, \delta[x - y]$ in equation (B.5b) to obtain,

$$\boldsymbol{v}\, \delta[\boldsymbol{v} - \boldsymbol{V}_i(t)] = \boldsymbol{V}_i\, \delta[\boldsymbol{v} - \boldsymbol{V}_i(t)]$$

Equation (B.5a) is the charge density of N_α point charges located at points $\boldsymbol{R}_i(t)$ and equation (B.5b) the sum of their individual fluxes.

The fields $\boldsymbol{E}^m(\boldsymbol{r}, t)$ and $\boldsymbol{B}^m(\boldsymbol{r}, t)$ acting between particles are highly variable along the inter-particle distance length scale $d_o = n_\alpha^{-1/3}$ and are responsible for short-range collisions[3]. In contrast, the fields originated by the collective influence of a large number of charges are shielded over longitudes longer than the Debye length scale $L_s \gtrsim \lambda_D$. The electrostatic approximation $\boldsymbol{E}^m(\boldsymbol{r}, t) = -\nabla\phi^m$ can be employed when charged particle currents are small and/or the magnetic field is essentially static compared with the plasma time scales.

The charge and current densities (B.5a) and the above Maxwell equations (B.4) for $\boldsymbol{E}^m(\boldsymbol{r}, t)$ and $\boldsymbol{B}^m(\boldsymbol{r}, t)$, together with the equations of motion (B.3) and the initial conditions for all plasma particles provide a complete and exhaustive—but impractical—description of a plasma. However, determination of the exact motion of all charged particles is obviously impossible. As in chapter 3 of volume 1 for the equilibrium neutral gases, we are mostly interested in statistical averages based on a tractable set of equations, whose solutions can be related to physically measurable macroscopic properties.

The first step is to take the time derivative of the probability distribution function f_α^m (B.1) of the α-particles as,

[3] The short-range collisions involve electromagnetic fields at the atomic and molecular level.

$$\frac{\partial f_\alpha^m}{\partial t} = \sum_\alpha \sum_{i=1}^{N_\alpha} \delta[v - V_i(t)] \frac{\partial}{\partial t} \delta[\boldsymbol{r} - \boldsymbol{R}_i(t)] + \delta[\boldsymbol{r} - \boldsymbol{R}_i(t)] \frac{\partial}{\partial t} \delta[v - V_i(t)] \quad \text{(B.6)}$$

Specifically, for one component[4] we have,

$$\frac{\partial}{\partial t} \delta[x - X_i] = \frac{\partial X_i}{\partial t} \frac{\partial}{\partial X_i} \delta[x - X_i] = \frac{\partial X_i}{\partial t} \left(-\frac{\partial}{\partial x} \delta[x - X_i] \right) = -V_{x,i} \frac{\partial}{\partial x} \delta[x - X_i]$$

using again $x\, \delta[x - y] = y\, \delta[x - y]$ to switch between Lagrangian to Eulerian velocities. We get for the first term,

$$\delta[v - V_i] \frac{\partial}{\partial t} \delta[\boldsymbol{r} - \boldsymbol{R}_i] = -\delta[v - V_i]\, V_i \cdot \nabla_{\mathbf{r}}\, \delta[\boldsymbol{r} - \boldsymbol{R}_i] = -\boldsymbol{v} \cdot \nabla_{\mathbf{r}}\, \delta[\boldsymbol{r} - \boldsymbol{R}_i]\, \delta[v - V_i]$$

where $\nabla_{\mathbf{r}}$ is the nabla operator in the Cartesian configuration space (x, y, z). Similarly, introducing the particle acceleration $\boldsymbol{a}_i = dV_i/dt$ given by the Lorenz force (B.3) and the nabla operator $\nabla_{\mathbf{v}}$ in the velocity components (v_x, v_y, v_z) the second term in equation (B.6) becomes,

$$\delta[\boldsymbol{r} - \boldsymbol{R}_i] \frac{\partial}{\partial t} \delta[v - V_i] = -\boldsymbol{a} \cdot \nabla_{\mathbf{v}}\, \delta[\boldsymbol{r} - \boldsymbol{R}_i]\, \delta[v - V_i]$$

where $\boldsymbol{a} = d\boldsymbol{r}/dt$ is the acceleration in continuum Eulerian coordinates. Finally introducing the definition (B.1) in equation (B.6) it reads,

$$\frac{\partial f_\alpha^m}{\partial t} + \boldsymbol{v} \cdot \nabla_{\mathbf{r}} f_\alpha^m + \frac{q_\alpha}{m_\alpha} [\boldsymbol{E}^m(\boldsymbol{r}, t) + \boldsymbol{v} \times \boldsymbol{B}^m(\boldsymbol{r}, t)] \cdot \nabla_{\mathbf{v}} f_\alpha^m = 0 \quad \text{(B.7)}$$

This last expression is called *Klimontovich equation* and mathematically it incorporates the motion of N_α particles in a partial differential equation for the velocity distribution function $f_\alpha^m(\boldsymbol{r}, \boldsymbol{v}, t)$ that is a function of seven continuous independent variables $(\boldsymbol{r}, \boldsymbol{v} = d\boldsymbol{r}/dt, t)$.

The expression (B.7) can be interpreted as a continuity equation for $f_\alpha^m(\boldsymbol{r}, \boldsymbol{v}, t)$ because[5] the fields $\boldsymbol{E}^m(\boldsymbol{r}, t)$ and $\boldsymbol{E}^m(\boldsymbol{r}, t)$ are independent of the velocity of particles,

$$\nabla_{\mathbf{v}} \cdot [\boldsymbol{E}^m + \boldsymbol{v} \times \boldsymbol{B}^m] = \cancelto{0}{\nabla_{\mathbf{v}} \cdot \boldsymbol{E}^m} + \nabla_{\mathbf{v}} \cdot (\boldsymbol{v} \times \boldsymbol{B}^m) = -\nabla_{\mathbf{v}} \cdot (\boldsymbol{B}^m \times \boldsymbol{v}) = -\boldsymbol{v} \cdot \cancelto{0}{(\nabla_{\mathbf{v}} \times \boldsymbol{B}^m)} = 0$$

Then $f_\alpha^m\, \nabla_{\mathbf{v}}\, \boldsymbol{a} = 0$ and we can write $\nabla_{\mathbf{v}} \cdot (f_\alpha^m\, \boldsymbol{a}) = \boldsymbol{a} \cdot \nabla_{\mathbf{v}} f_\alpha^m$ in equation (B.7) that now reads,

[4] We have made use of

$$\frac{\partial}{\partial x} \delta[x - y] = -\frac{\partial}{\partial y} \delta[x - y].$$

[5] We use the vector identity $\nabla_{\mathbf{v}} \cdot (g\, \boldsymbol{A}) = \boldsymbol{A} \cdot \nabla_{\mathbf{v}} g + g\, (\nabla_{\mathbf{v}} \cdot \boldsymbol{A})$ where $g(\boldsymbol{r}, \boldsymbol{v}, t)$ and $\boldsymbol{A}(\boldsymbol{r}, \boldsymbol{v}, t)$ are scalar and vector functions, respectively.

$$\frac{\partial f_\alpha^m}{\partial t} + \nabla_\mathbf{r} \cdot \left(f_\alpha^m \, \boldsymbol{v}\right) + \nabla_\mathbf{v} \cdot \left(f_\alpha^m \, \boldsymbol{a}\right) = 0 \tag{B.8}$$

Therefore, the Klimontovich equation (B.7) is equivalent to equation (B.8) which expresses the time conservation of the velocity distribution function $f_\alpha^m(\boldsymbol{r}, \boldsymbol{v}, t)$ (see equation A.4) in the six-dimensional phase space $(\boldsymbol{r}, \boldsymbol{v})$.

B.2 Phase space averaging

The exact description of a plasma at the microscopic scale provided by equations (B.1), (B.4), (B.5) and (B.7) is equivalent to equations (B.3), (B.5) and (B.4). However, equation (B.7) still considers the trajectories of N_α particles in the phase space and it is too detailed.

A simplified expression that governs the time evolution of a smooth velocity[6] distribution function $f_\alpha(\boldsymbol{r}, \boldsymbol{v}, t)$ can be rigorously derived from the Klimontovich equation (B.7) by *ensemble averaging*. This procedure essentially considers the mathematical average over a large number of virtual or imaginary copies of the system that are consistent with its macroscopic properties. The rapidly varying contribution over microscopic scales becomes smoothed out by the average of the velocity distribution function $f_\alpha^m(\boldsymbol{r}, \boldsymbol{v}, t)$ in the phase space.

However, using the fact that $\boldsymbol{r}$, $\boldsymbol{v}$ and t are independent variables, we will use instead a more intuitive procedure to average a given physical quantity $g^m(\boldsymbol{r}, \boldsymbol{v}, t)$ over the plasma microscopic scales. As we showed in volume 1 chapter 4, section 4.5.2, the *plasma frequency* $f_{p\alpha} = 1/T_\alpha$ provides the faster response to fluctuations of α-particles and,

$$\frac{1}{T_\alpha} \int_t^{t+T_\alpha} g^m(\boldsymbol{r}', \boldsymbol{v}', t') \, dt'$$

averages the contributions to g^m with characteristic times shorter than T_α. Additionally, in *ideal plasmas* a cube with Debye edge length contains $N_{D,\alpha} = n_\alpha \, \lambda_D^3 \gg 1$ a large number of particles (see section 4.5.3 in volume 1). Then, we can average over the intermediate length scale L_c,

$$d_\alpha = n_\alpha^{-1/3} \ll L_c \lesssim \lambda_D \tag{B.9}$$

that is much longer than the interparticle spacing d_o but below the Debye length[7]. Using the shorter response time T_α the associated range of speeds is,

$$d_\alpha \times f_{p,\alpha} \ll v_c = L_c / T_\alpha \lesssim \lambda_D \times f_{p\alpha} \tag{B.10}$$

[6] The velocity distribution function is equivalent to a probability distribution function, see chapter 2.

[7] Strictly speaking, the Debye length (equation (4.13) in volume 1) is only valid for Maxwellian distributions of electrons and ions. This is not the general case, so in its place we can consider λ_D as an *effective length scale* for electric field shielding along which the plasma quasineutrality is preserved.

The average of $g^m(\boldsymbol{r}, \boldsymbol{v}, t)$ over fast time scales $\tau_s > T_c$ within the six-dimensional cube

$$\Delta r \times \Delta v_c = L_c^3 \times v_c^3$$

located at the point $(\boldsymbol{r}, \boldsymbol{v})$ of the phase space is,

$$\langle g^m(\boldsymbol{r}, \boldsymbol{v}, t)\rangle = \frac{1}{T_\alpha}\int_t^{t+T_\alpha} dt' \frac{1}{L_c^3}\int_{\Delta r} d\boldsymbol{r}' \frac{1}{v_c^3}\int_{\Delta v} g^m(\boldsymbol{r}', \boldsymbol{v}', t')\, d\boldsymbol{v}' \tag{B.11}$$

where $d\boldsymbol{r}' = dx'\, dy'\, dz'$ and $d\boldsymbol{v}' = dv_x'\, dv_y'\, dv_z'$. The position and velocity of the phase space volume $\Delta r\, \Delta v$ simply are,

$$\langle \boldsymbol{r}\rangle = \frac{1}{L_c^3}\int_{\Delta r} \boldsymbol{r}'\, d\boldsymbol{r}' = \boldsymbol{r} \text{ and,} \quad \langle \boldsymbol{v}\rangle = \frac{1}{v_c^3}\int_{\Delta v_c} \boldsymbol{v}'\, d\boldsymbol{v}' = \boldsymbol{v}$$

Using equation (B.11) we can smooth the rapidly changing microscopic velocity distribution $f_\alpha^m(\boldsymbol{r}, \boldsymbol{v}, t)$ over longitudes $L_c \gg d_\alpha$ as,

$$\left\langle f_\alpha^m(\boldsymbol{r}, \boldsymbol{v}, t)\right\rangle = \frac{1}{L_c^3\, v_c^3} \times \left(\sum_{i=1}^{N_\alpha}\int_{\Delta r}\int_{\Delta v} \delta[\boldsymbol{r}' - \boldsymbol{R}_i(t)]\, \delta[\boldsymbol{v}' - \boldsymbol{V}_i(t)] d\boldsymbol{r}' d\boldsymbol{v}'\right) = f_\alpha(\boldsymbol{r}, \boldsymbol{v}, t)$$

Hence, $f_\alpha(\boldsymbol{r}, \boldsymbol{v}, t)\, d\boldsymbol{r}\, d\boldsymbol{v}$ represents the number of particles within the six--dimensional small phase space volume $\Delta r\, \Delta v_c \simeq d\boldsymbol{r}\, d\boldsymbol{v}$ centered at the point $(\boldsymbol{r}, \boldsymbol{v})$. Exchanging the order of integration in equations (B.2) and (B.11),

$$\int_{-\infty}^{+\infty} f_\alpha(\boldsymbol{r}, \boldsymbol{v}, t)\, d\boldsymbol{v}\, d\boldsymbol{r} = \frac{1}{L_c^3\, v_c^3}\int_{\Delta r} d\boldsymbol{r}' \int_{\Delta v} d\boldsymbol{v}' \left(\frac{1}{N_\alpha}\sum_{i=1}^{N_\alpha}\int_{-\infty}^{+\infty} \delta[\boldsymbol{r} - \boldsymbol{R}_i(t)]\, \delta[\boldsymbol{v} - \boldsymbol{V}_i(t)]\, d\boldsymbol{v}\, d\boldsymbol{r}\right) = N_\alpha$$

The averaged velocity distribution $f_\alpha(\boldsymbol{r}, \boldsymbol{v}, t)$ accounts for the properties of a smooth distribution for α-particles over longitudes $L_c \lesssim \lambda_D$, but that are large compared with the inter-particle spacing. Using the averaging procedure above discussed we can write,

$$f_\alpha^m(\boldsymbol{r}, \boldsymbol{v}, t) = f_\alpha(\boldsymbol{r}, \boldsymbol{v}, t) + \delta f_\alpha$$

with $\left\langle \delta f_\alpha \right\rangle = 0$ and δf_α corresponds to short range particle effects within the small volume $\Delta r\, \Delta v_c = L_c^3\, v_c^3$ of the phase space.

The electromagnetic fields are independent of the particle velocities and the Coulomb long-range interaction decays on a length scale that is long compared with the average distance d_α between particles. Equation (B.11) preserves the plasma collective effects averaging the microscopic electromagnetic fields over the length scale (B.9), that is longer than the interparticle spacing d_o but below the plasma Debye length scale.

$$\langle \boldsymbol{E}^m(\boldsymbol{r}, t)\rangle = \frac{1}{T_\alpha} \times \frac{1}{L_c^3} \times \int_t^{t+T_\alpha}\int_{\Delta r} \boldsymbol{E}^m(\boldsymbol{r}', t')\, d\boldsymbol{r}'\, dt' = \boldsymbol{E}(\boldsymbol{r}, t) \tag{B.12}$$

We also can write $\boldsymbol{E}^m(\boldsymbol{r}, t) = \boldsymbol{E}(\boldsymbol{r}, t) + \delta\boldsymbol{E}$, similarly $\boldsymbol{B}^m(\boldsymbol{r}, t) = \boldsymbol{B}(\boldsymbol{r}, t) + \delta\boldsymbol{B}$ and the averages of electromagnetic fields over short length scales are $\langle\delta\boldsymbol{E}\rangle = \langle\delta\boldsymbol{B}\rangle = 0$. Substituting into the microscopic Maxwell equations (B.4) and averaging the microscopic electromagnetic fields we obtain,

$$\begin{aligned} &\nabla \cdot \boldsymbol{E} = \frac{\rho_c}{\varepsilon_o}, \quad \nabla \times \boldsymbol{E} = -\frac{\partial \boldsymbol{B}}{\partial t} \\ &\nabla \cdot \boldsymbol{B} = 0, \quad \nabla \times \boldsymbol{B} = \mu_o\left(\boldsymbol{J}_c + \varepsilon_o \frac{\partial \boldsymbol{E}}{\partial t}\right) \end{aligned} \tag{B.13}$$

the mesoscopic field equations where the sources of the fields $\boldsymbol{E}(\boldsymbol{r}, t)$ and $\boldsymbol{B}(\boldsymbol{r}, t)$ are,

$$\rho_c(\boldsymbol{r}, t) = \sum_\alpha q_\alpha n_\alpha = \sum_\alpha q_\alpha \int_{-\infty}^{+\infty} f_\alpha \, d\boldsymbol{v} \tag{B.14a}$$

$$\boldsymbol{J}_c(\boldsymbol{r}, t) = \sum_\alpha n_\alpha q_\alpha \langle \boldsymbol{v}_\alpha \rangle = \sum_\alpha q_\alpha \int_{-\infty}^{+\infty} \boldsymbol{v} f_\alpha(\boldsymbol{r}, \boldsymbol{v}, t) \, d\boldsymbol{v} \tag{B.14b}$$

Finally, introducing f_α^m together with the fields $\boldsymbol{E}^m$ and $\boldsymbol{B}^m$ into the Klimontovich equation (B.7),

$$\begin{aligned} \frac{\partial}{\partial t}(f_\alpha + \delta f_\alpha) + \boldsymbol{v} \cdot \nabla_{\mathbf{r}} (f_\alpha + \delta f_\alpha) + \frac{q_\alpha}{m_\alpha}[(\boldsymbol{E} + \boldsymbol{v} \times \boldsymbol{B}) + (\delta\boldsymbol{E} + \boldsymbol{v} \times \delta\boldsymbol{B})] \\ \cdot \nabla_{\mathbf{v}} (f_\alpha + \delta f_\alpha) = 0 \end{aligned}$$

and averaging over the short time and length scales as above we derive the *plasma kinetic equation* for the smoothed $f_\alpha(\boldsymbol{r}, \boldsymbol{v}, t)$ velocity distribution function,

$$\frac{\partial f_\alpha}{\partial t} + \boldsymbol{v} \cdot \nabla_{\mathbf{r}} f_\alpha + \frac{q_\alpha}{m_\alpha}(\boldsymbol{E} + \boldsymbol{v} \times \boldsymbol{B}) \cdot \nabla_{\mathbf{v}} f_\alpha = -\frac{q_\alpha}{m_\alpha}\langle(\delta\boldsymbol{E} + \boldsymbol{v} \times \delta\boldsymbol{B}) \cdot \nabla_{\mathbf{r}} \delta f_\alpha\rangle \tag{B.15}$$

The left side of this partial differential equation (B.15) governs the evolution of the averaged probability distribution function $f_\alpha(\boldsymbol{r}, \boldsymbol{v}, t)$ for long times $\tau_s \gg T_\alpha$ and lengths $L_s \gtrsim \lambda_D$ in the phase space $(\boldsymbol{r}, \boldsymbol{v})$. The fields $\boldsymbol{E}(\boldsymbol{r}, t)$ and $\boldsymbol{B}(\boldsymbol{r}, t)$ in equation (B.15) satisfy the averaged Maxwell equations (B.13) for the charge $\rho_c(\boldsymbol{r}, t)$ and current $\boldsymbol{J}_c(\boldsymbol{r}, t)$ densities given by equations (B.14b) and (B.14a).

The fluctuating fields $\delta\boldsymbol{E}$ and $\delta\boldsymbol{B}$ within the average of the right side are given by equations (B.4) and (B.5) and are related with the correlation functions $\langle f_\alpha^m f_\beta^m \rangle$ (with $\alpha \neq \beta$) that can be connected with interactions of particle species across distances $L_c \lesssim \lambda_D$ below the Debye length. The derivation of correlation functions in relation to atomic and molecular encounters are beyond the scope of present work and can be found in the monograph [2].

However, following a more intuitive approach we can introduce a *collision operator* in equation (B.15) that needs to be specified,

$$C(f_\alpha) = \sum_{\beta,\alpha} C(f_\alpha, f_\beta) = -\frac{q_\alpha}{m_\alpha}\langle(\delta \boldsymbol{E} + \boldsymbol{v} \times \delta \boldsymbol{B}) \cdot \nabla_{\mathrm{r}}\, \delta f_\alpha\rangle$$

and accounts for the rate of change of the smoothed velocity distribution function $f_\alpha(\boldsymbol{r}, \boldsymbol{v}, t)$ by collisions with other plasma particles. Additionally, the equation of (B.15) can also be cast as a continuity equation,

$$\frac{Df_\alpha}{Dt} = \frac{\partial f_\alpha}{\partial t} + \boldsymbol{v} \cdot \nabla_{\mathrm{r}} f_\alpha + \boldsymbol{a} \cdot \nabla_{\mathrm{v}} f_\alpha = \sum_\beta C(f_\alpha, f_\beta) \qquad \text{(B.16)}$$

that expresses the rate of change experienced by $f_\alpha(\boldsymbol{r}, \boldsymbol{v}, t)$ in the seven-dimensional space $(\boldsymbol{r}, \boldsymbol{v}, t)$ due the electromagnetic fields and short-range particle interactions (collisions) between all plasma species.

Equation (B.16), usually denominated the *Boltzmann equation* and the averaged velocity distribution function $f_\alpha(\boldsymbol{r}, \boldsymbol{v}, t)$ are the starting point of chapter 2 where we also derive some approximate expressions for the collision operator. Comparing equation (B.16) with the time derivative (A.4), the collision operator C_α can be seen as a source–sink term which accounts for the time variation of particles within the volume $d\boldsymbol{r}\, d\boldsymbol{v}$ in the $(\boldsymbol{r}, \boldsymbol{v}, t)$ space.

When correlations (collisions) between discrete particles in equations (B.15) or (B.16) are negligible and the equation for time evolution of the averaged probability density becomes,

$$\frac{Df_\alpha}{Dt} = 0 \text{ and, } \quad \frac{\partial f_\alpha}{\partial t} + \boldsymbol{v} \cdot \nabla_{\mathrm{r}} f_\alpha + \frac{q_\alpha}{m_\alpha}(\mathcal{E} + \boldsymbol{v} \times \mathcal{B}) \cdot \nabla_{\mathrm{v}} f_\alpha = 0$$

This *collisionless* kinetic equation is called *Vlasov equation* and $f_\alpha(\boldsymbol{r}, \boldsymbol{v}, t)$, as well as any other function $H(f_\alpha)$ are constant in time in this case. However, the evolution of the plasma charged species is interrelated by electromagnetic fields of equation (B.13).

B.3 Commentaries and further reading

The derivation of the Klimontovich and Boltzmann equations in this appendix is essentially based on chapter 3 of the book [3]; a thorough and complete presentation of this theory can be found in the monograph [2]. The derivation of equation (B.16) based on the Liouville equation and the BBGKY[8] hierarchy describing a large number of interacting particles is discussed in chapter 4 of [3].

References

[1] Cohen-Tannoudji C, Diu B and Laloë F 1977 *Mécanique Quantique* (Paris: Hermann)

[2] Klimontovich Y L 1967 *The Statistical Theory of Non-equilibrium Processes in a Plasma* (Oxford: Pergamon)

[3] Nicholson D R 1983 *Introduction to Plasma Theory* 2nd edn (New York: Wiley)

[8] This is the acronym of Bogoliubov, Born, Green, Kirkwood and Yvon, the authors of this theory.

Lightning Source UK Ltd.
Milton Keynes UK
UKHW051254231220
375661UK00002B/32